EAI/Springer Innovations in Communication and Computing

Series editor
Imrich Chlamtac, European Alliance for Innovation, Gent, Belgium

Editor's Note

The impact of information technologies is creating a new world yet not fully understood. The extent and speed of economic, life style and social changes already perceived in everyday life is hard to estimate without understanding the technological driving forces behind it. This series presents contributed volumes featuring the latest research and development in the various information engineering technologies that play a key role in this process.

The range of topics, focusing primarily on communications and computing engineering include, but are not limited to, wireless networks; mobile communication; design and learning; gaming; interaction; e-health and pervasive healthcare; energy management; smart grids; internet of things; cognitive radio networks; computation; cloud computing; ubiquitous connectivity, and in mode general smart living, smart cities, Internet of Things and more. The series publishes a combination of expanded papers selected from hosted and sponsored European Alliance for Innovation (EAI) conferences that present cutting edge, global research as well as provide new perspectives on traditional related engineering fields. This content, complemented with open calls for contribution of book titles and individual chapters, together maintain Springer's and EAI's high standards of academic excellence. The audience for the books consists of researchers, industry professionals, advanced level students as well as practitioners in related fields of activity include information and communication specialists, security experts, economists, urban planners, doctors, and in general representatives in all those walks of life affected ad contributing to the information revolution.

About EAI

EAI is a grassroots member organization initiated through cooperation between businesses, public, private and government organizations to address the global challenges of Europe's future competitiveness and link the European Research community with its counterparts around the globe. EAI reaches out to hundreds of thousands of individual subscribers on all continents and collaborates with an institutional member base including Fortune 500 companies, government organizations, and educational institutions, provide a free research and innovation platform.

Through its open free membership model EAI promotes a new research and innovation culture based on collaboration, connectivity and recognition of excellence by community.

More information about this series at http://www.springer.com/series/15427

Fadi Al-Turjman
Editor

Smart Cities Performability, Cognition, & Security

Editor
Fadi Al-Turjman
Department of Computer Engineering
Antalya Bilim University
Antalya, Turkey

ISSN 2522-8595   ISSN 2522-8609   (electronic)
EAI/Springer Innovations in Communication and Computing
ISBN 978-3-030-14720-4   ISBN 978-3-030-14718-1   (eBook)
https://doi.org/10.1007/978-3-030-14718-1

This Springer imprint is published by the registered company Springer Nature Switzerland AG.
The registered company address is: Gewerbestrasse 11, 6330 Cham, Switzerland

To my wonderful family.

"You cannot connect the dots looking forward; you can only connect them looking backwards. So you have to trust that the dots will somehow connect in your future."
—*Steve Jobs*

Preface

We are living in an era where smart and cognitive solutions are becoming a global platform for the computation and interaction between humans as well as the machines while performing several critical tasks.

Performability, cognition, and security have been considered as a complementary package toward realizing the emerging smart cities paradigm. From this perspective, it is essential to understand the role of these three significant components which will provide a comprehensive vision for the worldwide smart city project in the near future.

No doubt that introducing such a new paradigm can come up with potential challenges in significant levels, especially in terms of the overall system performance, cognition, and security. It is also essential to consider the emerging intelligent applications for better lifestyle and more optimized solutions in our daily life.

The objective of this book is to overview existing smart cities applications while focusing on performability, cognition, and security issues. The main focus is on the smart design aspects that can help in realizing such paradigm in an efficient and secured way. The artificial intelligent (AI) techniques as well as new emerging technologies such as the Internet of Things (IoT) and the Smart-Cloud accompanied with critical evaluation metrics, constraints, and open research issues are included for discussion. This conceptual book, which is unique in the field, will assist researchers and professionals working in the area to better assess the proposed smart cities paradigms which have already started to appear in our societies.

Hope you enjoy it.

Fadi Al-Turjman

Contents

About the Editor

Fadi Al-Turjman is a professor at Antalya Bilim University, Turkey. He received his Ph.D. in computing science from Queen's University, Canada, in 2011. He is a leading authority in the areas of smart/cognitive, wireless, and mobile networks' architectures, protocols, deployments, and performance evaluation. His record spans over 200 publications in journals, conferences, patents, books, and book chapters, in addition to numerous keynotes and plenary talks at flagship venues. He has authored/edited more than 12 published books about cognition, security, and wireless sensor networks' deployments in smart environments with Taylor & Francis and Springer (top-tier publishers in the area). He is a recipient of several recognitions and best paper awards at top international conferences. He led a number of international symposia and workshops in flagship IEEE conferences. He is serving as the lead guest editor in several journals, including the *IET Wireless Sensor Systems* and *Sensors*, *MDPI Sensors*, and the Elsevier *Internet of Things*.

Chapter 1
An Effective Design for Polar Codes over Multipath Fading Channels

Jehad M. Hamamreh

1.1 Introduction

Thanks to the rapid advancements and great developments in both computing and communication technologies, smart cities are becoming not only a reality but also more popular and spreading day by day. In fact, the deployment of smart, cognitive cities is expected to be very dominant in many parts of the world in the near future due to their many advantages and merits represented by making life easier, faster, simpler, and safer [1–4].

Smart cities are composed of massive amount of smart, intelligent devices that have sensing, computing, actuating, and communication capabilities, designed in such a way that reduces human intervention through cognition and automation. These smart devices, commonly termed as Internet of Things (IoT) devices, are expected to dominate smart cities infrastructure. Among the many design requirements of IoT devices, especially for ultra-reliability and low-latency communication (URLLC)-based 5G services, low complexity with high reliability communication comes as a first key priority besides energy efficiency, latency, and security [5]. To meet this design goal, polar coding is proposed and adopted to be used in future 5G and beyond systems as a strong, attractive, and indispensable channel coding scheme [6, 7].

Polar codes, recently proposed by Arikan in [8], have driven enormous efforts by the wireless research community because of their provably capacity-achieving

J. M. Hamamreh (✉)
Department of Electrical-Electronics Engineering, Antalya International (Bilim) University, Antalya, Turkey
e-mail: jehad.hamamreh@antalya.edu.tr

© Springer Nature Switzerland AG 2020
F. Al-Turjman (ed.), *Smart Cities Performability, Cognition, & Security*,
EAI/Springer Innovations in Communication and Computing,
https://doi.org/10.1007/978-3-030-14718-1_1

property alongside its low complexity. Specifically, polar codes can achieve the capacity of the binary symmetric channel (BSC), which is equivalent to an additive white Gaussian noise (AWGN) channel with binary inputs generated by binary phase shift keying (BPSK) modulator. However, polar codes are not universal, which means different polar codes are constructed and generated according to the specified value of signal-to-noise ratio (SNR), defined as the design-SNR. This issue becomes even more problematic when the channel is fading due to having multipath (which is the case in most wireless channels) as it requires new ways and methods to construct new channel-specific polar codes.

In an attempt to make polar codes work in fading channels as it does in BSC, several works and research studies have been performed. In [9], a hierarchical, fading channels-tailored polar coding scheme that uses nested coding is presented. It was shown that the scheme is capacity achieving and it can be extended to fading channels with multiple, but finite number of states. In [10], polar codes are applied to wireless channels. A procedure for obtaining the Bhattacharyya parameters associated with AWGN and Rayleigh channels is presented. In [11], tracking lower and upper bounds on Bhattacharyya parameters of the bit subchannels was performed to construct good polar codes. In [12], the author analyzed polar coding strategies for the radio frequency (RF) channel with known channel state information (CSI) at both ends of the link and with known channel distribution information (CDI). Moreover, the investigation of polar coding for block-fading channel was performed in [13]. In [14], a polar coding scheme for fading channels was proposed. In particular, by observing the polarization of different binary symmetric channels over different fading blocks, each channel use, corresponding to a different polarization, is modeled as a binary erasure channel such that polar codes could be adopted to encode over blocks.

In [15], authors analyzed the general form of the extrinsic information transfer curve of polar codes viewed as multilevel codes with multistage decoding. Based on this analysis, they proposed a graphical design methodology to construct polar codes for inter-symbol interference channels. In [16], a simple method for the construction of polar codes for Rayleigh fading channel was presented. The subchannels induced by the polarizing transformation are modeled as multipath fading channels, and their diversity order and noise variance are tracked.

In [17], the polar codes are designed exclusively for block-fading channels. Specifically, the authors constructed polar codes tailored for block-fading channels by treating fading as a kind of polarization and then matching it with code polarization. The obtained codes are demonstrated to deliver considerable gain compared to conventional polar BICM schemes. In [18], the same authors provided an alternative design of polar codes for block-fading channels by treating the combination of modulation, fading, and coding as a single entity. This design is based on the fact that the bit channels are polarized not only by code, but also by fading and modulation. This observation enables constructing polar codes by mapping code polarization with modulation and fading polarization. The obtained codes adapt to the fluctuation of the channel.

In [19], the authors studied the problem of polar coding for non-coherent block-fading channels without considering any instantaneous channel state information. The scheme proposed by authors achieves the capacity of binary-input block-fading channels with only channel distribution information. In [20], the authors proposed a polar coding scheme for orthogonal frequency division multiplexing (OFDM) systems under multipath frequency selective fading channels. In this scheme [20], the codeword bits are permuted in such a way that the bits corresponding to the frozen bits are assigned to subcarriers causing frequent bit errors. This permutation can be considered as a kind of interleaving that can noticeably improve the polarization of the channel, thereby enhancing the bit error rate performance.

As seen from the literature, the previous studies on polar codes for fading channel have mainly focused on constructing new specific polar codes suitable to a particular fading channel. However, changing the polar codes construction based on the channel is not desirable in practice, since it would result in a continuous change and modification in the code construction based on the channel type, which is considered to be a cumbersome, complex, and inefficient process especially for IoT-based applications. Unlike the previous works, where new specific codes are constructed based on the channel, in this work, we propose a generic design solution, which enables us to use the same polar coding design, adopted in AWGN channel, for fading channels. The design neither causes any change in the encoder and decoder sides nor degrades the reliability performance. This is made possible through canceling the channel fading effect by using special channel-based transformations along with optimal power allocation,[1] so that a net, effective AWGN channel can be seen at the input of the successive cancelation decoder (SCD), whose simplicity and low complexity make it attractive to polar codes. In this work, a frequency selective fading channel is considered, which is the most common observed channel in broadband wireless systems.

The rest of the paper is organized as follows. System model is described in Sect. 1.2. The details of the developed design are revealed in Sect. 1.3. Then, simulation results are discussed in Sect. 1.4. Finally, conclusion and future works are drawn in Sect. 1.5.

Notations Vectors are denoted by bold-small letters, whereas matrices are denoted by bold-large letters. $\mathbf{I}$ is the $N \times N$ identity matrix. The convolution operator is indicated by $(*)$. The transpose and conjugate transpose are symbolized by $(\cdot)^T$ and $(\cdot)^H$, respectively.

---

[1]It should be stated that one obvious, common way to theoretically cancel the channel fading effect completely is to use channel inversion by means of applying zero forcing method at the transmitter[21]; however, this way is unfortunately impractical as it causes noise enhancement and a huge increase in the transmit power where it can go to infinity when the channel is in a deep fading situation.

1.2   System Model and Preliminaries

A single-input single-output (SISO) system, in which a transmitter (Tx) communicates with a receiver (Rx) as shown in Fig. 1.1, is assumed. All received signals
exhibit multipath slowly varying Rayleigh fading channels. The channel reciprocity
property is adopted, where the downlink channel can be estimated from the uplink
one, in a time division duplex (TDD) or hybrid systems (TDD with FDD) using
channel sounding [22]. Thus, the channel is assumed to be known at both the
transmitter and receiver sides [21, 23]. As per Fig. 1.1, a block of data bits denoted
by **u** is encoded using a binary polar coding scheme constructed at a fixed design-
SNR as depicted in Fig. 1.2, where N is the code length, K is the length of the
message sequence, and $J \subseteq N$, $|J| = K$ is the set of active indices known as
information bit indices, corresponding to the good channel observations [8, 24].
The remaining $N - K$ indices are called as frozen (inactive) bit indices. Here, N is

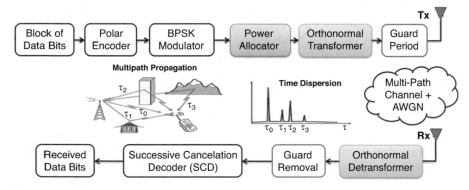

Fig. 1.1 Details of the optimal transceiver architecture for polar codes over a multipath frequency
selective channel

Fig. 1.2 Polar encoder
structure with $(N, K, J) =$
$(8, 5, \{2, 4, 6, 7, 8\})$

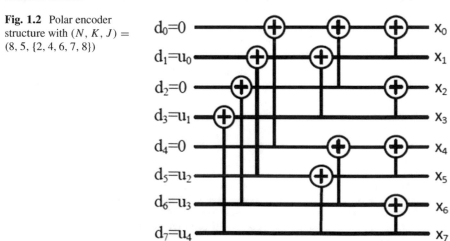

a power of 2 and we define $n = \log_2(N)$. For a (N, K, J) polar code, the generator matrix $\mathbf{G}$ is the n-fold Kronecker product of n copies of $\mathbf{F} = \begin{bmatrix} 1 & 0 \\ 1 & 1 \end{bmatrix}$. Therefore, given a message vector $\mathbf{u}$ of K information bits, a codeword $\mathbf{x}$ is generated as follows:

$$\mathbf{x} = \mathbf{u} \cdot \mathbf{G} = \mathbf{d} \cdot \mathbf{F}^{\otimes n}, \tag{1.1}$$

where $\mathbf{d}$ is a vector of N bits including information bits such that $\mathbf{d}_J = \mathbf{u}$, and $\mathbf{d}_{J^c} = \mathbf{0}$. The bits $\mathbf{d}_{J^c}$ are called as frozen bits, and are set to zero.

The codeword $\mathbf{x}$ is modulated using BPSK, resulting in a block of real modulated symbols, represented by

$$\mathbf{s} = \begin{bmatrix} s_0 & s_1 & \dots & s_{N-1} \end{bmatrix}^T \in \mathbb{C}^{[N \times 1]}. \tag{1.2}$$

In the proposed transmission scheme, each one of the real baseband modulated symbols s_i is assigned a specific power value e_i based on the quality of the corresponding experienced channel. For the N data symbols to be transmitted without interference over a time dispersive channel, we need N carrying orthogonal basis vectors, which can be taken in the proposed scheme from the column vectors of transformation matrix $\mathbf{V}$ (whose design details will be explained in the next coming section) given by

$$\mathbf{V} = \begin{bmatrix} \mathbf{v}_0 & \mathbf{v}_1 & \dots & \mathbf{v}_{N-1} \end{bmatrix} \in \mathbb{C}^{[N \times N]}. \tag{1.3}$$

Hence, $\mathbf{V}$ can be seen as the channel-based transformation matrix, which changes based on the user's channel. Also, each ith column vector in $\mathbf{V}$ can be expressed as

$$\mathbf{v_i} = \begin{bmatrix} v_0 & v_1 & \dots & v_{N-1} \end{bmatrix}^T \in \mathbb{C}^{[N \times 1]}. \tag{1.4}$$

After multiplying each symbol with its corresponding basis vector, we take the summation (superposition) of all the resulting weighted vectors to get a block of samples $\mathbf{g}$, referred to as one orthogonal transform division multiplexing (OTDM) symbol [25, 26]. The power allocation and transformation processes can mathematically be formulated as

$$\mathbf{g} = \sum_{i=0}^{N-1} e_i \cdot s_i \cdot \mathbf{v}_i \in \mathbb{C}^{[N \times 1]}, \tag{1.5}$$

which can further be simplified into a linear matrix representation form as follows:

$$\mathbf{g} = \mathbf{V}\mathbf{E}^{-1}\mathbf{s} \in \mathbb{C}^{[N \times 1]}, \tag{1.6}$$

where $\mathbf{E} = diag\begin{bmatrix} e_0 & e_1 & \dots & e_{N-1} \end{bmatrix}$ is a diagonal matrix, whose values represent the amount of power that is caused by each channel realization after transformation.

Moreover, to avoid the interference between consecutive adjacent blocks, known as inter block interference (IBI), zero-suffix padding, as a guard period interval with length equals to the length of the channel delay spread L, is appended to the end of each block. Zero-padding in our design can be understood as an off-transmission period, as is the case in ZP-OFDM. This results in saving power resources compared to CP-OFDM, since no energy is sent in the guard period. Additionally, extra unnecessary extension in guard period is avoided since the guard period length is set to be equal to the channel spread. After that, the OTDM symbol is sent through L-path slowly varying frequency selective fading channel with impulse response given as

$$\mathbf{h} = \begin{bmatrix} h_0 \; h_1 \; \ldots \; h_{L-1} \end{bmatrix}^T, \tag{1.7}$$

whose elements h_i are drawn from a complex Gaussian distribution function with zero mean and unity variance. The baseband received signal at the receiver can be given as

$$\mathbf{y} = \mathbf{h} * \mathbf{g} + \mathbf{z} \tag{1.8}$$

$$y_i = \sum_{l=0}^{L-1} h_l g_{(i-l)} + z_{(i)}, \tag{1.9}$$

where $\mathbf{y} = \begin{bmatrix} y_0 \; y_1 \; \ldots \; y_{N+L-1} \end{bmatrix}^T$ is the received block of one OTDM symbol and $\mathbf{z} \in C^{[(N+L-1)\times 1]}$ is the zero-mean real additive white Gaussian noise (AWGN). The previous convolution form can also be equivalently written in a linear algebraic matrix form, as

$$\mathbf{y} = \mathbf{H}\mathbf{g} + \mathbf{z} = \mathbf{H}\mathbf{V}\mathbf{E}^{-1}\mathbf{s} + \mathbf{z}, \tag{1.10}$$

where $\mathbf{H} \in \mathbb{C}^{[(N+L-1)\times N]}$ is the Toeplitz matrix form of the fading channel realizations between the transmitter and the receiver, given by

$$\mathbf{H} = \begin{bmatrix} h_0 & 0 & 0 & \ldots & 0 \\ h_1 & h_0 & 0 & \ldots & 0 \\ h_2 & h_1 & h_0 & \ldots & 0 \\ . & . & h_1 & \ldots & 0 \\ & & & \ldots & \\ . & . & . & \ldots & . \\ h_{L-1} & h_{L-2} & . & \ldots & . \\ 0 & h_{L-1} & h_{L-2} & \ldots & . \\ 0 & 0 & h_{L-1} & \ldots & . \\ \ddots & \ddots & \ddots & \ldots & . \\ 0 & 0 & 0 & \ldots & h_{L-1} \end{bmatrix}. \tag{1.11}$$

At the receiver, a channel-based transformation is performed on **y**, using a transformation matrix denoted by **U** that consists of multiple orthogonal basis vectors, which are optimally extracted from the channel to diagonalize the channel response. This process is then followed by successive cancelation decoding [8, 24] in the transform domain to decode the received data bits successfully. The process of extracting and using the matrix **U** will be discussed in the coming section.

1.3   Proposed Transceiver Design

The main goal of the proposed design is to compensate the effect of multipath frequency selective channel on the performance of polar codes, so that the required soft output symbols at the input of the successive cancelation decoder can be obtained without changing the polar code construction. This results in canceling the effect of fading and spreading caused by the multipath channel, leading to a very good performance (close to that of AWGN channel).

The main design steps are explained as follows:

- Assuming that **H** is available at both communication sides (transmitter and receiver) by means of channel sounding before the communication starts, and since the channel impulse response can be modeled as a Toeplitz matrix for block-based transmission, then both the transmitter and receiver can decompose **H** by applying any of the common decomposition methods (such as singular value decomposition (SVD), uniform channel decomposition (UCD), and geometric mean decomposition (GMD), etc.). For familiarity and simplicity, we choose SVD as the underlying method, thus **H** can equivalently be expressed as follows:

$$\mathbf{H} = \mathbf{UEV}^{H}. \tag{1.12}$$

- The transmitter takes the inverse of the diagonal matrix **E**, whose elements (singular values) represent the power spectrum of the parallel decomposed subchannels, and then uses it as a block-based power allocator (channel gain inverter or pre-equalizer).
- Before allocating the power to the modulated symbols, the transmitter classifies the N number of channel realizations into two group categories:

  (1) Invertible group, which can be used for symbol transmission as it corresponds to fading channel realizations that can be compensated by optimal power allocation without causing any increase in the transmit power (i.e., the average total transmit power remains unity).
  (2) Non-invertible group, which cannot be used for symbol transmission as it requires a huge amount of transmit power to compensate for the effect of deep fading channel realizations. The determination of which elements are

invertible and which are not is done by the transmitter via calculating the
sum of the diagonal elements of matrix $\mathbf{E}$ as follows:

$$P = \sum_{i=0}^{N-1} e_i = e_0 + e_1 + e_2 + \ldots + e_{N-1}. \tag{1.13}$$

Based on the value of P, the transmitter decides to select the first $\lfloor P \rfloor$
number of the channel realizations after transformation as carriers used for
symbol transmission, whereas the remaining ones $(N - \lfloor P \rfloor)$ are suppressed
(deactivated) and not used for symbol transmission. It should be noticed
that this selection mechanism can result in a slight degradation in the
effective throughput of the system due to the fact that the bad channel
realizations, which are responsible for degrading the reliability performance
of the system, are not used for data transmission, but rather suppressed and
excluded from being used for symbol transmission.

- The transmitter and receiver take the hermitian (conjugate transpose) of the right
  and left matrices, resulting from applying SVD decomposition on $\mathbf{H}$, i.e., $\mathbf{V}^H$
  and $\mathbf{U}$, to get $\mathbf{V}$ and $\mathbf{U}^H$, respectively. Then, use the matrices as transformer and
  de-transformer at the transmitter and receiver sides, respectively, to diagonalize
  the channel response and make it resilient to inter-symbol interference.

  It should be noted that the transform domain (analogous to the frequency
  domain) is obtained due to using $\mathbf{V}$ as an inverse fast Fourier transform (IFFT) at
  the transmitter and $\mathbf{U}^H$ as a fast Fourier transform (FFT) at the receiver. Without
  loss of generality, we first use the columns of $\mathbf{V}$ as orthogonal waveforms to
  carry the data symbols. The process of allocating channel gain-based power and
  assigning data symbols to waveforms and then summing them all can easily be
  expressed in a matrix form as in (1.6). When $\mathbf{g}$ passes through the channel and
  reaches the receiver side, the received OTDM block becomes as follows:

  $$\mathbf{y} = \mathbf{H}\mathbf{V}\mathbf{E}^{-1}\mathbf{s} + \mathbf{z} \in \mathbb{C}^{(N+L-1)\times 1}. \tag{1.14}$$

  As seen from the previous equation, since $\mathbf{H}$ can equivalently be written in
  terms of its SVD decomposition, then the pre-transformation matrix $\mathbf{V}$ used at
  the transmitter cancels the effect of the right part $\mathbf{V}^H$ of the channel since their
  multiplication results in an identity matrix $(\mathbf{I})$. Thus, the net received signal can
  be reformulated as

  $$\mathbf{y} = \mathbf{U}\mathbf{s} + \mathbf{z} \in \mathbb{C}^{(N+L-1)\times 1}. \tag{1.15}$$

- To remove the effect of the time dispersion brought by the channel spread caused
  by the left part of the channel $\mathbf{U} \in \mathbb{C}^{(N+L-1)\times N}$, the receiver needs to multiply
  the received signal by $\mathbf{U}^H$ as follows:

  $$\mathbf{U}^H\mathbf{y} = \mathbf{s} + \mathbf{U}^H\mathbf{z} = \mathbf{s} + \hat{\mathbf{z}} \in \mathbb{C}^{N\times 1}. \tag{1.16}$$

where $\hat{\mathbf{z}} = \mathbf{U}^H \mathbf{z}$, and because of the unitary nature of matrix $\mathbf{U}^H$, $\hat{\mathbf{z}}$ has the same statistics as $\mathbf{z}$.

It should also be noted that the vectors of $\mathbf{U}$ span not only the whole transmitted block time but also the following time reserved for zero-padding. Thus, $\mathbf{U}$ takes into account the spreading caused by the channel. The column vectors of $\mathbf{U}$ not only separate efficiently the transmitted symbols, leading to no interference between them, but also provide the best compromise between the useful energy and the noise collected in the zero-padding interval, leading to a maximum SNR increase for each transmitted block.

After multiplying by $\mathbf{U}^H$, the leakage energy of the signal due to channel spreading will be collected from the guard band optimally with minimal noise, thanks to the adaptive orthogonal waveforms, whose length at the receiver is equal to the received block length. Specifically, $\mathbf{U}^H$ maps the received block $\mathbf{y}$ from $\mathbb{C}^{(N+L-1) \times 1}$ to $\mathbb{C}^{(N) \times 1}$, resulting in accumulating the leaked energy and automatically removing the inserted guard period (ZP). Now, since the multipath channel is virtually converted to an equivalent AWGN channel as a result of applying the proposed signal processing scheme, a simple polar code decoder such as SCD can now be applied directly on the estimated block ($\mathbf{U}^H \mathbf{y}$).

It should be emphasized that since the transmitter performs channel gain inversion, the receiver no longer requires to perform equalization by $\mathbf{E}$, which otherwise would cause significant noise enhancement, data distribution changes, etc. and thus limit the achievable performance.

Moreover, the proposed design provides not only reliability close to that of AWGN channel without changing the simple polar coding structure, but also physical layer security for providing confidentiality against eavesdropping [23, 27–30] as an additional super advantage. Wireless physical security is delivered by the proposed scheme because of the fact that the transmitted data blocks are designed based on the channel of the legitimate receiver, which is naturally different than that of an eavesdropper's one, who is normally located several wavelength away from the legitimate receiver, and thus experiencing a different channel.

1.4 Simulation Results

We consider a block-based polar coding scheme with length $N = \{64, 128, 256\}$ and code rate $R = 0.5$. The polar codes are constructed at a fixed design-SNR equals to zero using Arikan's Bhattacharyya bounds [8] that is also explained in detail in [24]. BPSK modulation is then used, followed by optimal power allocation and orthonormal transformation as explained before in the previous section, and finally a guard period of length L is appended to the transmitted block. At the receiver, a de-transformation process is performed, followed directly by a simple and low complexity decoder, named as SCD [8, 24].

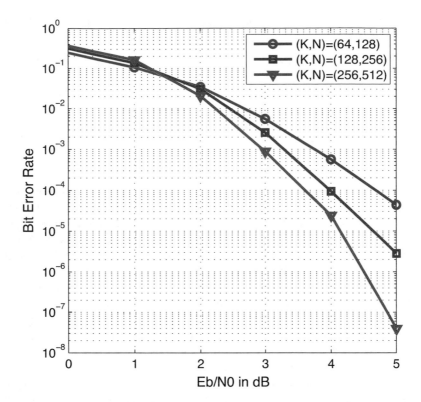

Fig. 1.3 BER performance of the proposed design over multipath channel

In the simulator, the channel is modeled as an independent and identically distributed (i.i.d.) block-fading, where channel coefficients are drawn from an i.i.d. Rayleigh fading distribution, at the beginning of each block transmission, and remain constant within one block, but change independently from one to another. The Rayleigh multipath fading channel has nine taps $L = 9$ with an exponential power delay profile given by

$$\mathbf{p}_{pdp} = [0.8407, \ 0, \ 0, \ 0.1332, \ 0, \ 0.0168, \ 0.0067, 0, \ 0.0027] \ \text{mW}, \quad (1.17)$$

where the distribution of the amplitude of each non-zero tap is assumed to be Rayleigh as commonly used in the literature.

Figure 1.3 presents the bit error rate (BER) performance versus E_b/N_0 (where Eb is the bit energy and N_0 is the power spectral density of the noise) of polar codes using the proposed transceiver architecture. On the other hand, Fig. 1.4 presents the frame error rate (FER) performance versus E_b/N_0 of the proposed design at different block lengths.

It is shown from the both figures that the achieved performance over a multipath fading channel is almost as same as that obtained over an AWGN channel[24].

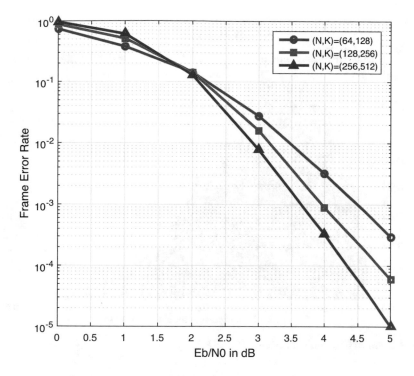

Fig. 1.4 BER performance of the proposed design over multipath channel

The gain is attained as a result of canceling the dispersion and fading effects of the channel by performing proper transformation and optimal power allocation alongside channel selection according to the channel behavior in such a way that the total response of the system is diagonalized and becomes similar to that of an AWGN channel. This means that the inter-symbol interference between data symbols s_i within one transmission block is totally removed.

Moreover, the way the interference leakage in the guard period is collected in the presence of noise, using $\mathbf{U}^H$, is optimal. Additionally, it is clear from the BER performance that as the block length increases, the performance gets improved, which is exactly similar to what happens in the case of an AWGN channel [24].

It should be stated that the achieved desirable reliability performance (in terms of BER/FER) of the proposed scheme that makes polar codes designed for AWGN [24] work over multipath fading dispersive channels with the same reliability as that of AWGN comes at a cost. This cost is related to a slight random degradation (loss) in the maximum data rate that can be transmitted over the multipath channel as explained in the aforementioned section. Therefore, it would be insightful to further understand and quantify the amount of data rate performance loss incurred by the proposed practical design. To clearly show and visualize this loss, we conduct a

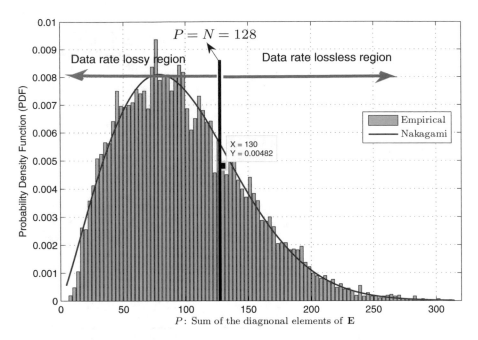

Fig. 1.5 Probability density function (PDF) of P

simulation experiment by which we generate 10,000 random multipath channel realizations whose power delay profile values are given as in (1.17). For each channel realization, whose length N is assumed to be equal to 128 samples, we decompose the Toeplitz matrix form of the multipath channel using SVD method to obtain the diagonal matrix $\mathbf{E}$ as explained in the aforementioned section. Once we have $\mathbf{E}$, we can then calculate P, which is the sum of the diagonal elements of $\mathbf{E}$. Since for each iteration we have one value for P, then we would have 10,000 values for P given the number of realizations we have which are 10,000 as well. For the obtained 10,000 values of P, we plot the probability density function (PDF) and cumulative distribution function (CDF) of P as shown in Figs. 1.5 and 1.6, respectively. It can be noticed from the figures that the PDF and CDF curves of the experimental values of P fit to Nakagami distribution. Also, it is shown that the whole distribution region can be divided into two regions: (1) data rate lossless region, whose sum power is greater than the number of channel samples (i.e., $P > N$); and (2) data rate lossy region, whose sum power is less than the number of channel samples (i.e., $P < N$).

Accordingly, in the first region, we have fully invertible subchannels, whereas in the second region, we have partially invertible subchannels. Therefore, we would have a certain data rate loss in the lossy region as the value of P is less than N and thus we cannot send data over all subchannels.

Furthermore, from Fig. 1.6, we can observe that the probability of being in the lossy region is around 75%, which is quite huge percentage, resulting in the need to

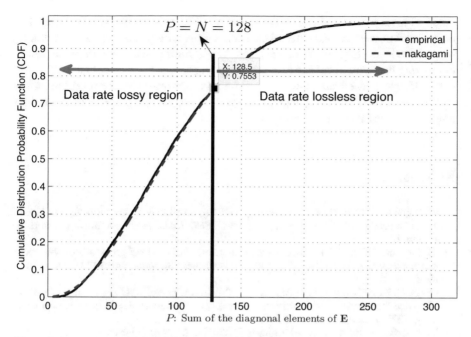

Fig. 1.6 Cumulative distribution function (CDF) of P

activate a certain number of subchannels according to the value of P. The number of deactivated subchannels (NDC) can be calculated as $NDC = N - \lfloor P \rfloor$, which increases as we move towards the left of the CDF curve, but with lower probability. This would result in limiting the amount of maximum data rate at which we can have a performance equal to that of an AWGN channel without changing the polar code design whatsoever. However, it should be pointed out that the deactivated subchannels can be utilized to do some other useful functionalities such as reducing interference, peak to average power ratio, and providing further physical layer secrecy [22].

1.5   Conclusion

In this work, we have proposed a novel solution that enables using the same polar coding design, used in AWGN channel, for multipath frequency selective fading channels without causing any change in the encoder and decoder sides while maintaining the same reliability performance. This has been made possible by canceling the channel fading effect by using a pre-transformation with optimal power allocation at the transmitter and a post-transformation at the receiver, so that a net AWGN channel could be seen at the input of the successive cancelation decoder. Future work will include the extension of the proposed design to other practical modulations such as M-QAM and M-PSK. Moreover, to limit the level of the transmit power and make the scheme compatible with practical power amplifiers, new methods are required to be designed. These future methods will be dedicated to utilizing the symbol subcarriers corresponding to low subchannel gains (i.e., the ones which are deactivated and not used for data transmission) for useful purposes and functionalities such as shaping the spectrum or reducing the peak to average power ratio (PAPR) of the transmit waveform.

References

1. Al-Turjman, F., Ever, E., & Zahmatkesh, H. (2018). Small cells in the forthcoming 5G/IoT: Traffic modelling and deployment overview. *IEEE Communications Surveys Tutorials, PP*(99), 1.
2. Al-Turjman, F. M. (2018). Modelling green femtocells in smart-grids. *Mobile Networks and Applications, 23*(4), 940–955. [Online]. Available: https://doi.org/10.1007/s11036-017-0963-1
3. Hasan, M. Z., Al-Turjman, F., & Al-Rizzo, H. (2018). Analysis of cross-layer design of quality-of-service forward geographic wireless sensor network routing strategies in green internet of things. *IEEE Access, 6,* 20,371–20,389.
4. Deebak, B. D., Ever, E., & Al-Turjman, F. (2018). Analyzing enhanced real-time uplink scheduling algorithm in 3GPP LTE-advanced networks using multimedia systems. *Transactions on Emerging Telecommunications Technologies, 29*(10), e3443.
5. Hamamreh, J. M., Ankarali, Z. E., & Arslan, H. (2018). CP-less OFDM with alignment signals for enhancing spectral efficiency, reducing latency, and improving PHY security of 5G services. *IEEE Access, 6,* 63,649–63,663.
6. Hui, D., Sandberg, S., Blankenship, Y., Andersson, M., & Grosjean, L. (2018). Channel coding in 5G new radio: A tutorial overview and performance comparison with 4G LTE. *IEEE Vehicular Technology Magazine, 13*(4), 60–69.
7. Mohammadi, M. S., Collings, I. B., & Zhang, Q. (2017). Simple hybrid ARQ schemes based on systematic polar codes for IoT applications. *IEEE Communications Letters, 21*(5), 975–978.
8. Arikan, E. (2009). Channel polarization: A method for constructing capacity-achieving codes for symmetric binary-input memoryless channels. *IEEE Transactions on Information Theory, 55*(7), 3051–3073.
9. Si, H., Koyluoglu, O. O., & Vishwanath, S. (2014). Polar coding for fading channels: Binary and exponential channel cases. *IEEE Transactions on Communications, 62*(8), 2638–2650.
10. Shi, P., Tang, W., Zhao, S., & Wang, B. (2012). Performance of polar codes on wireless communication channels. In *2012 IEEE 14th International Conference on Communication Technology (ICCT)* (pp. 1134–1138). Piscataway: IEEE.
11. Zhang, Y., Liu, A., Pan, K., Gong, C., & Yang, S. (2014). A practical construction method for polar codes. *IEEE Communications Letters, 18*(11), 1871–1874.
12. Bravo-Santos, A. (2013). Polar codes for the Rayleigh fading channel. *IEEE Communications Letters, 17*(12), 2352–2355.
13. Boutros, J. J., & Biglieri, E. (2013). Polarization of quasi-static fading channels. In *2013 IEEE International Symposium on Information Theory* (pp. 769–773). Piscataway: IEEE.
14. Islam, M. K. & Liu, R. (2013). Polar coding for fading channel. In *2013 International Conference on Information Science and Technology (ICIST)* (pp. 1096–1098). Piscataway: IEEE.
15. Fayyaz, U. U. & Barry, J. R. (2014). Polar code design for intersymbol interference channels. In *Global Communications Conference (GLOBECOM), 2014 IEEE* (pp. 2357–2362). Piscataway: IEEE.

16. Trifonov, P. (2015). Design of polar codes for Rayleigh fading channel. In *2015 International Symposium on Wireless Communication Systems (ISWCS)* (pp. 331–335). Piscataway: IEEE.
17. Liu, S., Hong, Y., & Viterbo, E. (2017). Polar codes for block fading channels. In *Wireless Communications and Networking Conference Workshops (WCNCW), 2017 IEEE* (pp. 1–6). Piscataway: IEEE.
18. Liu, S., Hong, Y., & Viterbo, E. (2017). Adaptive polar coding with high order modulation for block fading channels. In *2017 IEEE International Conference on Communications Workshops (ICC Workshops)* (pp. 755–760). Piscataway: IEEE.
19. Zheng, M., Chen, W., & Ling, C. (2018). Polar coding for noncoherent block fading channels. In *2018 10th International Conference on Wireless Communications and Signal Processing (WCSP)* (pp. 1–5). Piscataway: IEEE.
20. Oda, M. & Saba, T. (2018). Polar coding with enhanced channel polarization under frequency selective fading channels. In *2018 IEEE International Conference on Communications (ICC)* (pp. 1–6). Piscataway: IEEE.
21. Deekshith, P. K., & Sahasranand, K. R. (2017). Polar codes over fading channels with power and delay constraints. In *2017 International Symposium on Wireless Communication Systems (ISWCS)* (pp. 37–42). Piscataway: IEEE.
22. Hamamreh, J. M., Basar, E., & Arslan, H. (2017). OFDM-subcarrier index selection for enhancing security and reliability of 5G URLLC services. *IEEE Access, 5,* 25,863–25,875.
23. Hamamreh, J. M., Furqan, H. M., & Arslan, H. (2018). Classifications and applications of physical layer security techniques for confidentiality: A comprehensive survey. *IEEE Communications Surveys Tutorials, 2018,* 1.
24. Vangala, H., Viterbo, E., & Hong, Y. (2015). A comparative study of polar code constructions for the AWGN channel. *CoRR, abs/1501.02473.* [Online]. Available: http://www.polarcodes.com/
25. Hamamreh, J. M., & Arslan, H. (2017). Secure orthogonal transform division multiplexing (OTDM) waveform for 5G and beyond. *IEEE Communications Letters, 21*(5), 1191–1194.
26. Hamamreh, J. M., & Arslan, H. (2017). Time-frequency characteristics and PAPR reduction of OTDM waveform for 5G and beyond. In *2017 10th International Conference on Electrical and Electronics Engineering (ELECO)* (pp. 681–685). Piscataway: IEEE.
27. Guvenkaya, E., Hamamreh, J. M., & Arslan, H. (2017). On physical-layer concepts and metrics in secure signal transmission. *Physical Communication, 25,* 14–25.
28. Furqan, H. M., Hamamreh, J. M., & Arslan, H. (2017). Enhancing physical layer security of OFDM-based systems using channel shortening. In *2017 IEEE 28th Annual International Symposium on Personal, Indoor, and Mobile Radio Communications (PIMRC)* (pp. 8–13). Piscataway: IEEE.
29. Hamamreh, J. M., Furqan, H. M., & Arslan, H. (2017). Secure pre-coding and post-coding for OFDM systems along with hardware implementation. In *Wireless Communications and Mobile Computing Conference (IWCMC), 2017 13th International* (pp. 1338–1343). Piscataway: IEEE.
30. Hamamreh, J. M., & Arslan, H. (2018). Joint PHY/MAC layer security design using ARQ with MRC and null-space independent PAPR-aware artificial noise in SISO systems. *IEEE Transactions on Wireless Communications, 17*(9), 6190–6204.

Chapter 2
LearningCity: Knowledge Generation for Smart Cities

Dimitrios Amaxilatis, Georgios Mylonas, Evangelos Theodoridis, Luis Diez, and Katerina Deligiannidou

2.1 Introduction

Smart cities have slowly been turning from a vision of the future to a tangible item, through the efforts of numerous research projects, technological start-ups, and enterprises, combined with the recent advancements in informatics and communications. It is currently a very active field research-wise, with a lot of works dedicated to developing prototype applications and integrating existing systems, in order to make this move from a vision to reality.

While there is still a wealth of ongoing activity in the field, in terms of technologies competing as candidates for mainstream adoption, at least we have a number of applications slowly emerging and taking shape inside smart city instances. For example, much buzz surrounds the smart city IoT testbed and experimentation concept, like in the case of SmartSantander [30]. Another example is the utilization of open data portals in smart cities, like CKAN [14], an open source solution provided by a worldwide community, and Socrata [32], an enterprise solution backed by an IT company. Additionally, protocols like MQTT and communication technologies like LoRa or NB-IoT are used in recent smart city research projects to provide real-time communication with the deployed infrastructure, and progress towards becoming Internet standards.

However, essential answers are to be found revolving around a central question: What do we do with all of these data collected, and how do we make sense out

D. Amaxilatis · G. Mylonas (✉) · E. Theodoridis · K. Deligiannidou
Computer Technology Institute and Press "Diophantus", Patras, Greece
e-mail: amaxilat@cti.gr; mylonasg@cti.gr; theodori@cti.gr; kdeligian@cti.gr

L. Diez
University of Cantabria, Santander, Spain
e-mail: ldiez@tlmat.unican.es

© Springer Nature Switzerland AG 2020
F. Al-Turjman (ed.), *Smart Cities Performability, Cognition, & Security*,
EAI/Springer Innovations in Communication and Computing,
https://doi.org/10.1007/978-3-030-14718-1_2

of them by extracting knowledge, i.e., something actually useful, going beyond a technology demonstrator? Related to the previous question, we also need to find a way to provide usefulness to citizens, involving them in the smart city lifecycle, and engaging them in the city shaping. By creating more "useful" information out of raw sensors, or other kind of data representing observations of the urban environment, better ways to discover data streams and easy ways to utilize the information will create significant acceleration in the smart city ecosystems. For example, certain events generate data reported by the city sensing infrastructure, but are, more often than not, missing an appropriate description. Consider the case of a traffic jam inside the city center. It generates sensed values in terms of vehicles speed, noise, and gas concentration. In addition, in most cases, multiple devices or services, while missing useful correlations in the data streams, report such values.

We believe that adding data annotations to smart city data, through machine-learning or crowdsourcing mechanisms, can help in revealing a huge hidden potential in smart cities. In this sense, one key aspect is how to traverse through this sea of smart city data in order to decide where and what to look for before adding such annotations. In addition, we also have to analyze correlations from findings after having processed this data, in order to uncover the hidden information inside them.

In this work, we discuss the design and implementation of JAMAiCA (Jubatus Api MAChine Annotation), a system for aiding smart city data annotation through classification and anomaly detection. On the one hand, it aims to simplify the creation of more automated forms of knowledge from data streams, while on the other hand it serves as a substrate for crowdsourcing data annotations via a large community of contributors that participate in the knowledge creation process. We strongly believe that communities like data scientists, decision makers, and citizens should get involved in deployments of future Internet systems, for them to be practical and useful.

In order to validate our approach, in this work we include a number of illustrative use cases for which we provide some evaluation results. These use cases utilize a combination of data sources that provide insights to the actual conditions in the center of the city of Santander, like parking spots, traffic intensity, and weather sensors, along with the load in wireless telecommunication networks. Based on our preliminary analysis, our findings show some interesting correlations between the aforementioned datasets that could be of interest to city planners, local authorities, and citizen groups.

Moreover, the system present here was designed, implemented, employed, and evaluated inside the context of OrganiCity[1] ecosystem. OrganiCity, as a smart city technology ecosystem, aims to engage people in the development of future smart cities, bringing together three European cities: Aarhus (Denmark), London (UK), and Santander (Spain). Co-creation with citizens is its fundamental principle, i.e., defining novel scenarios for more people-centric applications inside smart cities,

---

[1]Co-creating digital solutions to city challenges, https://organicity.eu/.

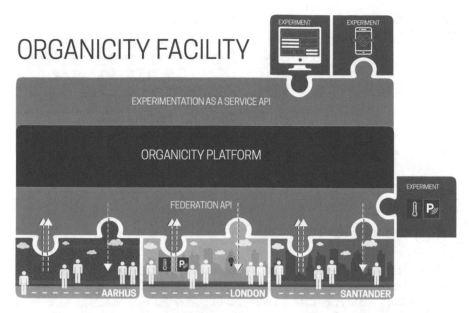

Fig. 2.1 A high-level view of OrganiCity as a smart city platform. The platform offers a Federation API through which 3 European cities (Santander, Aarhus, and London) with smart city infrastructure are "federated." This federation enables experimenters who utilize the experimentation-as-a-service API of the platform to see these 3 cities through a single interface. Additional options are available through smartphone experimentation and additional cities' federation with the platform

exploiting the IoT technologies, heterogeneous data sources, and using enablers for urban service creation and IoT technologies. Figure 2.1 provides an overview of OrganiCity. In short, the project aims to provide an experimentation-as-a-service (EaaS) platform [8], i.e., it is designed to make data streams from diverse sources inside a smart city available to various "consumers," like IoT experimenters, SMEs, municipalities, etc. At the same time, it aims to enable the participatory engagement of communities in co-creating urban knowledge. This is done by means of end-user applications that provide meaningful representations of the produced smart city data, and "tools" that will allow these end users to make their own contributions.

Regarding the structure of this work, we first report on previous related work, and continue with a discussion on challenges associated with knowledge creation in smart cities. We then present a small set of use cases to highlight how our system relates to this vision. We continue with a presentation of our design and system architecture, complemented with a description of our current implementation and some preliminary results we have produced so far. Finally, we summarize the main contributions brought about by the JAMAiCA system, and highlight some aspects that will be tackled by exploiting the annotation system (Table 2.1).

Table 2.1 List of
abbreviations used in the text

IoT	Internet of Things
API	Application Programming Interface
CKAN	Comprehensive Knowledge Archive Network
MQTT	Message Queuing Telemetry Transport
LoRa	Long Range
NB-IoT	Narrowband IoT
JAMAiCA	Jubatus Api MAChine Annotation
EaaS	Experimentation-as-a-Service
DL	Deep Learning
NGSI	Next Generation Service Interface
UDO	Urban Data Observatory

2.2   Previous Work

Although there have been a number of recent studies and applications aiming to combine human and machine intelligence, research in this field is still at its infancy stage. In [18], authors present a vision on the potential of combining patterns of human and machine intelligence, identifying three possible patterns: sequential, parallel, and interactive. Moreover, in [9] authors present a crowd-programming platform that integrates machine and human based computations. Their system integrates mechanisms for challenging tasks like human task scheduling, quality control, latency due to human behavior, etc.

A key element in most of the approaches is the use of a taxonomy or ontology, which are ubiquitous in organizing information. They represent formal structures to represent entities, to organize entities to categories, to express their relations, and to map data objects to abstract concepts expressing meanings, entities, events, etc. Most of the modern social networking applications and online collaborative tools are heavily relying on an underlying taxonomy. Building and curating a taxonomy is a challenging task that requires deep knowledge of a specific domain and the corresponding data characteristics, usually performed by a small group of experts of the application domain. In contrast to proprietary domains, folksonomies are quite popular in online applications and they are organically created by their users. Such taxonomies usually have weaknesses like double entries, misclassified tags, entries with typos or ambiguities in the classes, and so on.

In [12], the authors propose a workflow that creates a taxonomy from collective efforts of crowd workers. In particular, the taxonomization task breaks down into manageable units of work and an algorithm coordinates the work that involves mapping to categories, identifying the best mappings and judge the relevance of the associated categories. Although there exist taxonomies and tagging of objects with keywords of the taxonomy, the problem is that there is no common agreement about the semantics of a tagging, and thus every system uses a different representation. In [22], an effort for the development of a common tagging ontology with semantic web technologies is described.

Designing and developing smart cities is a concept that has drawn tremendous attention from the public and the private sector. Each one of the scientific disciplines like urban engineering, computer science, sociology, and economics provides unique perspectives on making cities more efficient. In most of these cases, multidisciplinary approaches are required to tackle complex problems. Projects employing machine and crowdsourced learning techniques started to take shape the last years.

SONYC [10] is an example of a project with a very well-defined use case, employing machine-learning algorithms to classify acoustic readings into various types of noise encountered inside an urban environment. It is a very interesting approach, with similarities to our vision of providing a generic substrate to simplify the process of knowledge extraction and data annotation contributions. Moreover, learning from the crowds by using the crowdsourced labels in supervised learning tasks in a reliable and meaningful way is investigated in [29, 38].

A large number of projects are trying to leverage modern information and communication technologies, like IoT/future Internet and the semantic web, in order to build novel smart city services and applications. An example is the SmartSantander project [30], which has developed one of the largest future Internet infrastructures globally, located at the center of the city of Santander in Spain. A well-established citywide IoT experimentation platform moved testbeds from labs to the real world and offers experimentation functionality, both with static and mobile deployed IoT devices, together with smartphones of volunteers inside the urban areas. van Kranenburg et al. [36] discuss the SOCIOTAL EU project, which attempted to tackle co-creation aspects inside smart cities. Another example is CitySDK [13] that tries to harmonize APIs across cities and provide guidelines about how information should be modeled, propose ways to exchange data, and how services and applications should be designed and developed. The project benefits from semantic web technologies, and focuses on application domains like citizen participation, mobility, and tourism.

CityPulse [27] introduces a framework for real-time semantic annotation of streaming IoT and social media data to support dynamic integration into the Web. The framework employs a knowledge-based approach for the representation of the data streams. It also introduces a lightweight semantic model to represent IoT data streams, built on top of well-known models, such as TimeLine Ontology, PROV-O, SSN, and Event Ontology. In terms of creating high-level concepts from the large amount of data produced, another similar approach has been carried out in [17]. The latter approach introduces a methodology to automatically create a semantic ontology, without requiring preliminary training data, using an extended k-means clustering method and applying a statistical model to extract and link relevant concepts from the raw sensor data. In [11] the authors propose principles for semantic modeling of city data, while in [28] the authors propose a technique to extract hidden structures and relations between multiple IoT data streams. The method employs latent Dirichlet allocation (LDA) on top of meaningful abstractions that describe the numerical data in human understandable terms.

In [19] the authors present a IoT deployment enhanced with a machine learning and semantic reasoning layers on top. Moreover, in [25] they explore by surveying the application of deep learning (DL) techniques on IoT and smart city streams and try to surface challenges, limitations, and opportunities. Aggarwal et al. [2] discuss the IoT field from a data-centric perspective. PortoLivingLab [31] is a smart city deployment project supporting multi-source sensing from IoT deployments and crowdsourcing in order to achieve city-scale sensing focusing on weather, environment, public transport, and people flows. Finally, they present a set of use cases that provide key insights into the status of city of Porto, Portugal. In [24], the authors discuss smart mobility scenarios that are representative for big cities, especially in China.

Regarding the infrastructure described and used for our evaluation, [30] introduce aspects of the utilized infrastructure, while [33] discuss issues and lessons from the deployment and operation of such a large IoT smart city infrastructure. In practice, through our evaluation we have detected issues in the operation and continuity of data, which as mentioned in [33] are aspects which could be serviced by systems like the one discussed here. The use case of parking inside the city is presented in detail in [23]. The issue of overall data quality in IoT is discussed in [21].

Regarding other recent work in IoT and smart cities from a security and reliability perspective, in [3] an agile framework regarding authentication, confidentiality, and integrity is discussed, while [4] propose a context-sensitive seamless identity provisioning (CSIP) framework for the IoT in the healthcare domain. Al-Turjman et al. [5] discusses the unreliable nature of communication networks used in real-world IoT platforms. Alabady and Al-Turjman [6] and Alabady et al. [7] present aspects related to security and reliability of communication in the IoT domain.

The work presented in this article acts as both an end-user tool and a service for other applications to extend the data annotation functionality of a wider smart city ecosystem. Currently it has been employed and tested in the OrganiCity smart city technology infrastructure, especially data stream discovery services like Urban Data Observatory (UDO).[2] An earlier version of the work discussed here was presented in [15], while a detailed discussion of some of the findings regarding experimentation in OrganiCity is included in [8]. In the latter work, we provided a detailed discussion on the design and implementation, along with additional related results.

2.3   Data Annotation in Smart Cities: Challenges

In this section, we briefly discuss a set of key challenges regarding data annotation in smart cities. We can partition them in two fundamental objectives:

- enable a more engaging and secure experience for citizens/contributors, and
- produce a more meaningful results/observations from the system side.

---

[2]https://docs.organicity.eu/UrbanDataObservatory/.

Privacy and overall security issues are a central challenge in the context discussed here. Consider the case of a volunteer taking noise level measurements along his daily commute, or being tasked to add annotation contributions by a smart city system based on proximity to certain events. Even in such simple scenarios, anonymization techniques should be used to ensure that neither personal data nor interactions are revealed.

Another important issue is the correlation of different types of smart city data that can potentially point to the same event, in other words, how to facilitate knowledge extraction through such data. We currently have data produced by IoT infrastructure installed inside city centers. However, there is relatively small research focus on discovering relations between these data. For instance, we may analyze if noise level readings are related to a live concert event, or can be attributed to another event produced by a specific situation (e.g., traffic jam) taking place somewhere inside the city.

Moreover, there is the issue regarding the nature of data available in smart city data repositories, being data inserted by humans or IoT infrastructures. Both sources can be unreliable, or even malicious. With respect to sensing infrastructure, we also have the issue of the hardware malfunctions, as well as spatiotemporal effects on the data produced. In most cases, the hardware utilized aims for large-scale deployments, thus being not so accurate or having calibration issues. Additionally, environmental conditions, e.g., excessive temperature or humidity, may have an effect on the sensitivity of the sensing parts. In this regard, one key aspect is how to produce data annotation based on such an infrastructure, which can function with a varying degree of credibility during a single day. Reputation mechanisms are an example of measures that can aid in this direction, either human or machine-based, in order to filter out less reliable data sources.

The issue of end-user engagement with respect to data annotation and knowledge extraction is, in our opinion, another major challenge. We also think that user contribution is twofold: end users can contribute to a smart city system by adding data annotations, but also contribute data through incentivization or gamification. Although most current crowdsourcing platforms utilize a desktop or web interface, the crowdsourcing of data annotations does not have to be limited to that. It can be also performed through smartphones and be incorporated to the user's everyday life. The interaction of end users through such a tool could help in relating in a more personal way and maintaining the interest in participating. Moreover, annotation of events or sensed results could be more interactive and focus at users, or even user groups, near the actual space of the event in question.

Smart city facilities usually integrate a large number of data sources of various types sharing observations for environment, air quality, traffic, transport, social events, and so on. These data sources might be static (they are not streaming data and have a fixed value until they are updated by an offline process) or might be dynamic (streaming data constantly). Building a taxonomy on this multi-thematic environment is not straightforward, since some tags subcategories might be shared between different types of data sources, while others might be orthogonal. In addition, as the dynamic data sources have a temporal dimension, annotations might characterize the overall behavior, and observations, of data sources or observations falling into a specific time interval. Furthermore, as data sources might be mobile (e.g., an IoT device on a bus or a smartphone) an annotation might characterize a specific location inside the city within a specific time interval. Embedding these spatiotemporal characteristics in the taxonomy introduces new requirements and extensions to traditional methods. Standardization bodies, like W3 Web annotation data model and protocols, do not cover sufficiently these requirements.

Finally, implementing machine-learning algorithms suited to smart city data and real-time processing is another major challenge. Handling citywide data introduces additional complexity, especially when considering relations between different data types and sensing devices. Current mobile devices have enough processing power to handle a broad set of use cases, especially when dealing with data from integrated sensors (e.g., [34] uses on-device processing to classify urban noise sources). This could also be utilized as a means to enhance privacy, since processing would be performed locally, without requiring sensitive data to be uploaded to the cloud.

2.3.1   Use Cases

We now proceed with a set of characteristic use cases, aiming to highlight our vision of the annotations system, and to provide insights to tackle the aforementioned challenges.

- *IoT sensors to create better running and biking routes*: This use case utilizes mobile and smartphone/smartwatch sensors to monitor environmental parameters so as to infer better routes for running and biking in terms of healthy environmental conditions. Parameters that could be sensed include air quality, noise pollution, pollen concentration, condition of roads, etc. Machine-learning techniques could be used to identify anomalies in the sensed data, such as high pollutant or particle concentrations, or locations with high noise levels. Alerts regarding such events could be sent by the system to participating end users to quantify or validate such data through annotations. Another use of data annotation contributions could relate to the sentiments of participants for their surroundings.
- *The soundtrack of the city*: The concept is to create the aural and noise level maps of cities. This includes the use of the microphones of smartphones to record noise or distinctive sounds of the urban landscape. Participants could use data annotations to pinpoint street musicians, sounds from birds or other animals and their location, or sounds from public spaces like train, bus stations, or city halls, etc. Machine-learning techniques could be used to generate general classifications that could subsequently be made more specific by end users providing additional data annotations. Users could also add descriptions and their sentiments towards places and sounds.

- *Smart city event correlations*: A diverse smart city IoT infrastructure could "record" the same event from different aspects; a traffic jam could take place at a certain point in time (traffic data) while creating certain side effects, such as noise from car horns or engines (noise data), unusual levels of pollution (air quality data), etc. Since this kind of data is being fed to the system with similar spatiotemporal characteristics, such anomalies can be detected and correlated on a first level and then be validated by end users to define additional correlations.

2.4 Architecture

As outlined previously, OrganiCity federates existing smart city infrastructures, integrating urban data sets and services. Federated resources are exposed, in this context, through a unified experimentation service and a central context broker[16]. Based on this existing architecture, our data annotation service, JAMAiCA, is designed to operate over the updates provided by the context broker, in order to provide additional knowledge, increasing their value and usefulness. JAMAiCA is capable of consuming, processing, and annotating individual data points to produce temporal annotations or nearby measurements to generate spatial annotations.

Figure 2.2 presents the integration of the JAMAiCA service with the OrganiCity infrastructure. It also presents the two main software components and the other building blocks employed in order to generate and store the annotations for the IoT data received.

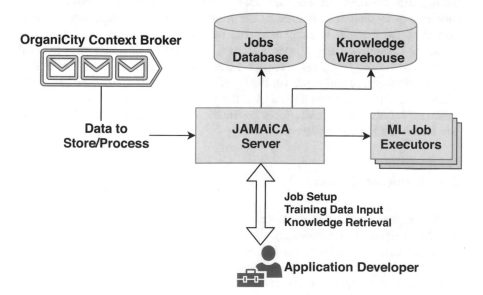

Fig. 2.2 Integration of the JAMAiCA service with the OrganiCity infrastructure. Interested application developers can utilize the datasets available from OrganiCity, by using the interfaces of the system to setup, configure, and monitor the execution of machine-learning jobs

The core module, the JAMAiCA server, offers both an API and a web interface for users to setup, configure, and monitor their machine-learning jobs. The configuration of each job is stored in the Jobs Database module. Once a job is setup, the JAMAiCA server subscribes for data updates to the context broker and sets up a proper executor for the machine-learning job. Each ML executor is configured based on the job configuration and fed with the training data provided by the user. When a new sensor measurement is received it is passed to the respective executor to be analyzed.

The Knowledge Warehouse is responsible for maintaining a directory of all possible annotations in the form of **tags**. Tags are simple indicators of the annotated parameter, similar to the way tagging is performed in photos in social networks, or the use of hashtags in social status updates. **Tag domains** are created as collections of tags (e.g., `high`, `normal`, and `low`) with a similar contextual meaning. Tag domains can be generic as those mentioned before or more application specific (e.g., the tag "`contains a traffic light`" for images). Users of the system can either select one of the tag domains already available or create their own specifically for their application. The outcomes of the analysis from the machine-learning executors are also stored in the Knowledge Warehouse with additional notes that can be numeric or text values. These notes can be user comments or a value that describes the abnormality of an observation, or a confidence indicator for the classification.

The JAMAiCA server is capable of performing both anomaly detection and classification jobs over the data streams formed by the updates from the context broker. In both cases, after the annotation jobs are added to the system, the initial training data are submitted and the back-end is trained. Annotation begins and for each data point examined the results are stored to the Knowledge Warehouse. Regarding the machine-learning executors themselves, the system is agnostic of the actual machine-learning frameworks, as it is capable of using multiple external services for the job. This gives us the flexibility to experiment with various machine-learning frameworks and expandability to easily provide extra functionality in the future. In our case, we evaluated during the development of the tool two distinct solutions: Jubatus and JavaML. More information on the operation of JAMAiCA is provided in the following section.

2.5 Implementation

In this section, we discuss the technologies used for the implementation of the JAMAiCA Server and Knowledge Warehouse, together with details regarding data communication and the available end-user interfaces. Regarding the implementation of the system, it is openly available on GitHub[3] via OrganiCity's repository, along with user guides and examples.

---

[3]https://github.com/OrganicityEu/JAMAiCA.

2.5.1   Communication and Frameworks Used

Regarding communication, the provision of data to our system is done either directly or through an NGSI context broker [26]. Additional options like ActiveMQ or MQTT message queues can be implemented and then be added to the system. For our main use case, JAMAiCA uses a context query, provided during the creation of the machine-learning job, to register for updates on the main OrganiCity context broker. This query acts as a set of selection parameters for the devices and sensors the job is interested in. In OrganiCity, the FIWARE Orion context broker is used. After a subscription is established, the context broker uses POST HTTP requests to notify our system of the newly received data following the NGSI specification. We also offer users the option to manually send data to our system via a similar HTTP POST request to a per job auto-generated endpoint. The format of the HTTP body needs to be formatted according to NGSI specification for the sake of simplicity.

Both the JAMAiCA server and Knowledge Warehouse are implemented using Java and the Spring Boot framework [35]. Spring Boot is Spring's convention-over-configuration solution for creating stand-alone, production-grade Spring-based applications, as it simplifies the bootstrapping and development stages. It eases the process of exposing components, such as REST services, independently and offers useful tools for running in production, database initialization, and environment specific configuration files and collecting metrics.

In our case, we implemented both interfaces as RESTful web services. The API of the JAMAiCA server offers methods for handling primary HTTP requests (post, get, put, and delete) that correspond to CRUD (create, read, update, and delete) operations on the *jobs database*, respectively. As a result, it allows experimenters to add and manage annotation jobs through their applications. A machine-learning job can be either an anomaly detection or a classification process. Since both jobs require initial training data, additional methods that allow training a machine-learning instance for an existing job are available.

2.5.2   Machine-Learning Frameworks

In order to perform the analysis of the data, we used the Jubatus Distributed Online Machine-Learning Framework [20] and JavaML [1]. In general, in the smart city plane it is usually not practical to use conventional approaches for data analysis by storing or analyzing all data as batch-processing jobs. Instead, our system processes data in online manner to achieve high throughput and low latency by using multiple frameworks and executors for online and distributed machine learning. It also processes all data in memory, and focuses on the actual operations for data analysis to update its machine-learning executors instantaneously just after receiving and analyzing the data. For each annotation process, we deploy a dedicated instance based on Jubatus or JavaML and feed it with the provided training data. Our service

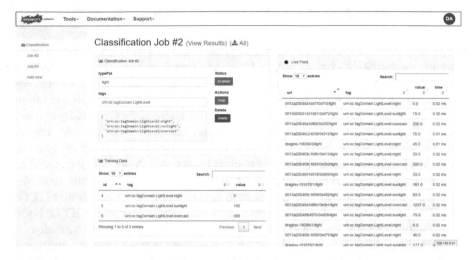

Fig. 2.3 An instance of the user interface of JAMAiCA for configuring and monitoring classification jobs. A basic classification job is set up in this case, related to light levels, where we see on the left part the description and training data for the job, while on the right there is a live feed with new values being classified

communicates with each instance using a wrapper with a common interface. This setup allows us to horizontally scale the machine-learning infrastructure on demand. An example of the graphical user interface of the system can be seen in Fig. 2.3.

2.5.3   Knowledge Warehouse

The **Knowledge Warehouse** uses Neo4j [37] to maintain the taxonomies generated by the **tags** and **tag domains**. Neo4j is a graph database that leverages data relationships and helps us build an intelligence around the entities stored in our system and the relationships between them. By traversing the relationships between **tags** that comprise a tag domain, we can easily create suggestions for the appropriate tags a new annotation application may use. Also, the relationships between annotations can be used to extract higher knowledge for the cities or the users of the system, especially when augmented with location and time metadata in order to identify events or situations that arise inside the cities. For example, an application can query for streets of the city where high atmospheric pollution and low vehicle speed are detected (indicating a possible traffic jam) and advise drivers to use alternative routes or means of transport, when combined with information about the local subway timetables.

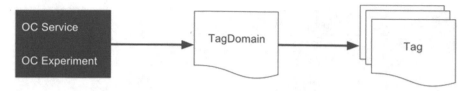

Fig. 2.4 OrganiCity services, experiments, tag domains, and tags

The underlying data model of the **Knowledge Warehouse** and the relations between its entities are depicted in Fig. 2.4:

- **Tags** represent the actual annotation labels to be used by users of the system.
- **Tag Domains** represent collections of tags. Usually a tag domain is associated with a service and/or an experiment specifying which tag domains they will use.
- **Services** represent utility/urban services. An example of a service might be garbage collection, or noise monitoring. The basic usage of service entities is the organization and discovery of tag collections (e.g., what **tags** are usually used for characterizing the noise level sensors).
- *Experiments* are created by users of the system.

In addition to the entities presented above, the following concepts come to glue together the annotation schema with the OrganiCity facility entities:

- **Annotations** are relationships between the **assets** of the OrganiCity and a **tag**.
- **Assets** are entities inside the OrganiCity facility that can be annotated. The assets are not stored in the internal database of the system but referenced by the added annotations.

An example of how assets are annotated inside the **Knowledge Warehouse** is presented in Fig. 2.5. In the figure, two assets of OrganiCity (the red circles) are annotated with two different **tags** (the blue circles) that belong to a single **tag domain** that describes the traffic conditions in the city.

2.5.4 End-User Interfaces and Integration with Existing Tools

Both modules presented above do not provide dedicated user interfaces for OrganiCity end users to add/validate or vote for available annotations. To handle this aspect, we provide interaction between users and **JAMAiCA** based on OrganiCity tools like the Urban Data Observatory (UDO[4]), a smartphone experimentation application

---

[4]OrganiCity's Data Observatory, https://observatory.organicity.eu/.

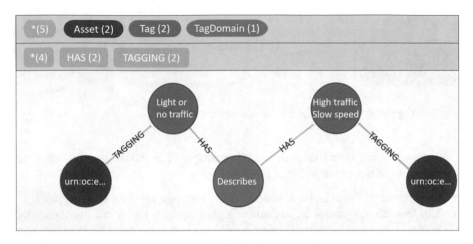

Fig. 2.5 Example of data annotations and the relationships created using the Neo4j server web interface

developed for OrganiCity (Sensing on the Go[5]), or other applications developed by OrganiCity Experimenters themselves. The Urban Data Observatory facilitates the search among the assets of OrganiCity, and users can use it to identify assets relevant to their interests. This is achieved by allowing for easy discovery and filtering of assets, based on geolocation, types, metadata, and even reputation or recommendations scores based on the experience, reliability, and opinions of other users. The JAMAiCA information presented above is offered to users by a bidirectional interaction with annotations allowing for:

- Looking up annotations of particular assets.
- Filtering assets based on annotations already stored in the JAMAiCA.
- Requesting to users to validate existing tags generated by JAMAiCA.
- Addition or deletion of tags for a particular asset on demand.

An example of how the annotation information is displayed on the UDO is available in Fig. 2.6.

The JAMAiCA annotations are also integrated into the Assets Discovery Service of OrganiCity, which is the back-end that allows the UDO to perform the data filtering. This is especially important since it provides end users with the ability to filter and search assets based on the stored tags. To provide this information on the Asset Discovery Service, annotation information is pushed to the Asset Directory Service every time they are created, updated, or deleted. However, this interface is not aimed at replacing the JAMAiCA API when retrieving the complete annotation for an specific asset, since the information stored on the Asset Discovery Service are in an aggregated and limited format.

---

[5]https://play.google.com/store/apps/details?id=eu.organicity.set.app.

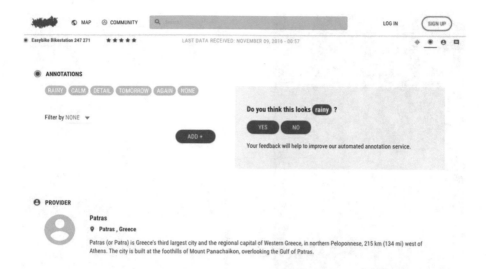

Fig. 2.6 The annotations module interface on the UDO. The end users are presented with the existing information and suggestions on how to improve data and add new aspects to the existing ones

As an additional interface for adding and collecting annotations, we have developed a specific view on the "Sensing on the Go" smartphone application that allows users to add and modify asset annotations as part of their experiment. This view loads a map view of the nearby area based on the user's location together with the available assets in their vicinity. On top of these pre-existing assets, participants of experiments can add annotations by simply selecting the appropriate **tag**. Experimenters can retrieve the generated annotations from the system and associate them with any data that their users have also collected from the phone sensors. The view for adding annotations from this application is available in Fig. 2.7.

An additional end-user interface is available through integration with the Tinkerspace platform. Tinkerspace[6] is an online tool for building simple smartphone "applications" that is part of the OrganiCity toolset offered to end-user communities. A number of functionality "blocks" are available, upon which users define their own rules for processing and I/O. In order to support Data Annotation in Tinkerspace, we have provided a new version of existing blocks to support annotation. To showcase the operation of the blocks we designed, we used the visual programming interface provided by Tinkerspace to generate our own application for annotating using smartphones. The interface of the Tinkerspace application can be seen in Fig. 2.7.

---

[6]http://www.tinkerspace.se/.

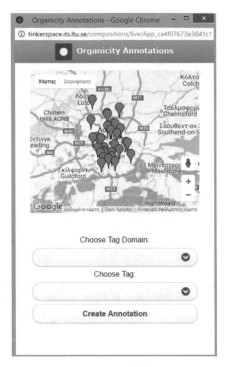

Fig. 2.7 Screenshots from the smartphone interface of data annotation-related software: Sensing on the Go app (left) and Tinkerspace interface (right). On the smartphone app, and as part of an experiment running on top of OrganiCity, developers can add options for creating annotations during the experimentation process. Such annotations are added by end users/volunteers who offer to execute such experiment on their Android smartphones while moving through the city. On Tinkerspace, the displayed interface is an option that can be added to an application for the system, to insert annotations

2.6 Results and Discussion

In this section, after having introduced our design and implementation, we proceed to discuss some illustrative cases that showcase both the usefulness of our system and the capacity of extracting knowledge out of smart city data. We base our evaluation on data produced by the smart city infrastructure available at Santander, Spain.

We have focused our analysis on the following data sources:

- Parking spots sensors: A number of sensors are installed under the actual parking spots at the center of Santander, continuously monitoring the available parking spots.
- Traffic intensity sensors: A number of sensors are installed at big traffic junctions, mostly located at the outer parts of the city, monitoring and counting the number

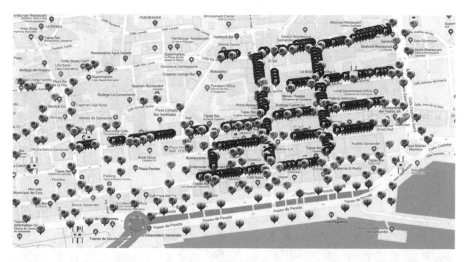

Fig. 2.8 The downtown area in Santander where the majority of the IoT nodes which generate our data sets are installed. It is roughly a 800 by 400 m area in the center of the city, containing parking, noise, weather, traffic, and e-field sensors

of vehicles. The readings are extrapolated to 1 hour, and are measured in vehicles per hour.

- Weathers sensors: Weather stations are scattered around the city, producing readings that categorize the type of weather. Values range from 0 to 11, based on the type of weather ("0" meaning sunny weather, "3" is cloudy, "7" is heavy rain, and "11" is hail).
- Electric-field sensors: The e-field sensors measure the electromagnetic field in downlink communications (base-station to users) and for all operators together, for the telecommunications bands used in Europe (2G-900 MHz, 2G-1800 MHz, 4G-1800 MHz, and 3G-2100 MHz). It is an indirect measure of the activity inside the city center.

For our evaluation purposes, we selected (see Fig. 2.8) a subset of the Smart-Santander infrastructure.[7] The area we selected is located at the center of the city and contains the vast majority of the parking sensors in the city, apart for the other types of sensors. It is located inside the main cultural and commercial district of the city, containing lots of offices, services, banks, shops, and restaurants, i.e., it is one of the busiest areas in Santander day and year-round. Examples of deployed devices are depicted in Fig. 2.9. Regarding the time in which the measurements were produced, data were generated during 2017, with the majority of the data examined here generated during the second half of 2017. This is due to the fact that this was the period with the smallest number of disruptions in data continuity, i.e., with less gaps in the datasets examined.

---

[7]For a map view of the whole installed infrastructure, please visit http://maps.smartsantander.eu/.

Fig. 2.9 Some examples of IoT devices deployed inside the center of Santander. Both stationary and mobile IoT nodes are used, installed on lamp posts, on top of buses, buried beneath the soil in public parks, or on the walls of buildings at the streets of Santander's center

Figure 2.10 provides us with a base understanding of how the parking spots in question are used by the citizens of Santander throughout the days we analyze. As we can see in the figure, the number of the available parking spots typically peeks around 3:00 in the morning and reaches the lowest points twice during each day, once around 10:00 in the morning (office hours) and around 17:00 in the afternoon when the stores are again open and people return to their work or visit the commercial district of the city. Another important characteristic we can identify in the data of this figure is that the number of the available parking spots in the area on Saturday mornings is higher than the rest of the days in the morning (96 vs 80 available spots). Similarly, on Sunday mornings we can see that people come to the area a bit earlier than on the rest of the days (around 9:00) and occupy more parking spots (40 vs 47 available spots) almost until the evening, when the spots are again free.

Figure 2.11 depicts the available parking spots at the center of Santander for a given time period, while Fig. 2.12 depicts, in more detail, the values of e-field for a given location in the same area and available parking spots for a specific week. In

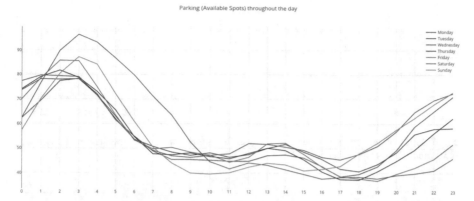

Fig. 2.10 Available parking spots at the center of Santander for each day of the week between August 15 and December 31, 2017

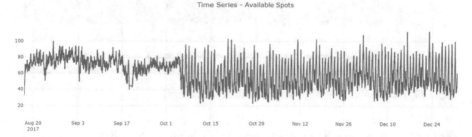

Fig. 2.11 Available parking spots at the center of Santander between August 15 and December 31, 2017

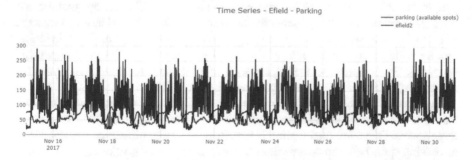

Fig. 2.12 Available parking spots at the center of Santander and e-field measurements for the area containing the parking spots, between November 15 and December 1, 2017

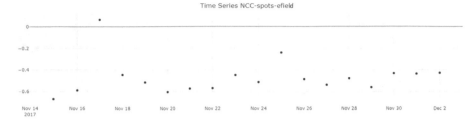

Fig. 2.13 Pearson correlation per day for parking spots and e-field measurements, between November 15 and December 1, 2017

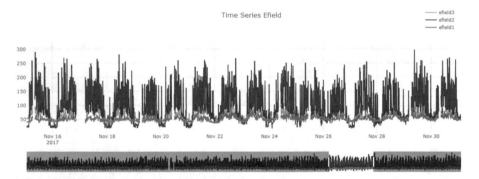

Fig. 2.14 E-field measurements from 3 locations inside Santander

many cases, there is a strong inverse correlation between the e-field values and the number of available parking spots at the center, as displayed in Fig. 2.13.

Regarding Fig. 2.14, we showcase the e-field values from 3 locations in the center. Overall, it seems that e-field values follow similar patterns, and the sensor that we consider for our study in the previous figures (*efield2*) is the one that records the higher values, i.e., it represents the busiest area. There seems to be an average e-field value that when it starts to change rapidly, parking spots are also declining rapidly within a limited amount of time, and go from 80 to 50 available parking spots quickly. It seems that the variability of e-field could serve as an indicator for creating a prediction for when the parking will start filling up.

Furthermore, throughout the year there are in some cases anomalies in parking values and e-field values that need context, e.g., during specific days we noticed large spikes. One additional finding is that there is a large difference in availability of parking spots between summer and winter periods; in fact, values almost double in winter. This is most probably due to the fact that there are several schools and university buildings close to this area, and as soon as the main activities for the school year begin, there is a very noticeable difference in parking spots availability.

One other interesting finding from our graphs is that, for this particular area with its specific characteristics in mind, it is quite possible that e-field readings can, to a certain extent, substitute parking sensors. In an application scenario which is based

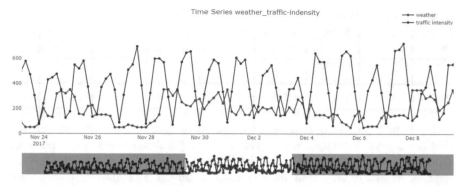

Time Series weather_traffic-indensity

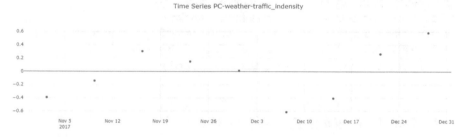

Fig. 2.15 Traffic intensity together with weather conditions during 2 weeks in November–December 2017. Even though there are several days in which there was rain (higher values mean worse weather or levels of rain), the values measured for traffic do not seem to be affected much

Fig. 2.16 Correlation between weather and traffic, showing results that on average there is weak correlation between the two parameters

on the overall availability of parking spaces inside the center, the e-field readings can provide a good approximation of the general picture. In the case where there are specific parking spots of a special type, e.g., parking spaces reserved for people with disabilities, then the approximation does not hold as well.

Moving on to traffic, when talking about the values from the sensors, since most sensors measure in/out rate to the city from the suburbs and overall rest of the world, it seems that rain does not affect the intensity of traffic. In the same week, there is no big irregularity during days when it is raining compared to ones with good weather. This is probably due to the fact that traffic intensity is mainly measured at the main entrances and exits for the city, and not inside the center, where traffic jams could be created more easily due to the weather conditions. Again, long-term measurements show in this case that traffic is half in summer compared to winter. Figure 2.15 shows the measurements for traffic intensity against weather conditions during a 2-week time period, while Fig. 2.16 shows the correlation between the 2 parameters. On average the Pearson correlation coefficient for November and December 2017 was 0.016960 and −0.137907, respectively.

In terms of adding annotations to smart city datasets, through this preliminary evaluation we can claim that there exist a number of scenarios where such annotations could provide missing context to the data. As such an example, anomalies can be detected in traffic intensity, which are not explained by other parameters, e.g., weather. In those cases, citizens can probably add annotations that could explain the situation to a certain extent.

Furthermore, anomaly detection can be used to detect periods where the infrastructure is down, and exclude those values from further analysis that could change average values, etc. In general, data anomalies detection can help in identifying hardware malfunctions or failures in specific points. Since there exist clear patterns that are observed throughout the year in all of the datasets we examined here, reporting abnormal values could possibly imply, e.g., battery depletion or failure. In the specific case of the e-field measurements, in certain cases, data anomalies were identified that indicated sensors sending measurements with zero values for a large part of the day. This was probably due to the fact that their batteries were beginning to fail, and when recharged through solar or power flowing only for certain hours, the batteries could not last as long as they did previously.

Moreover, irregularities can be identified in the specific case of the parking sensors, where there are periods, e.g., when the variability is very small considered to the rest of the days. This could be used to identify days where there are roadworks taking place, which was actually the case in some periods at the center of Santander.

2.7   Conclusions and Future Work

Research on smart cities is starting to produce tangible results, introducing new possibilities for citywide services and applications, as well as accelerating the adoption of new technologies. However, we believe that there is still work to be done with respect to establishing workflows and tools in order to produce results with a more tangible impact on citizens' lives. In this context, one of the directions the research community has recently taken is to combine the existing smart city datasets within machine-learning frameworks in more "clever" ways than in the past.

In this article, we presented a solution that validates and enriches data produced by smart city infrastructures. We believe that this kind of processing is critical for IoT sensor data to become something more than simple datasets, i.e., a useful and reliable data source to facilitate the development of future city services. We presented a service capable of automatically detecting erroneous or unexpected sensor data using machine-learning algorithms, classifying them and detecting real-world events and situations (in the form of annotation tags) based on provided training data sets.

We have included some examples of analysis of smart city data based on our system, based on parking, noise levels, traffic intensity, mobile network, and weather data, produced inside the city center of Santander. Again, we would like to emphasize that our intention with respect to our evaluation is to showcase the hidden potential of smart city data, which can be revealed given the right tools. Since we are not data scientists, city planners, or sociologists, our interpretation of the data presented in this work is bound to be limited by our own scientific background. Having this in mind, our evaluation revealed some interesting findings that could point to novel ways on designing future smart city infrastructure, or how to utilize the combination of data sources that have not been explored yet.

To achieve the next level of smart city data analysis, we believe that citizen participation is of critical importance. We plan on focusing on interactive interfaces that make it easy for users to augment or confirm the automated annotations generated from our system. Moreover, we believe that in order to further increase the participation rate and interest of the citizens, various methods for incentives and gamification should be assessed, either in the context of rewards from the city to the citizens or in the form of specific benefits/findings that city services could offer to the citizens or local authorities.

Acknowledgements This work has been partially supported by the EU research project OrganiCity, under contract H2020-645198.

References

1. Abeel, T., Peer, Y. V. D., & Saeys, Y. (2009). Java-ML: A machine learning library. *Journal of Machine Learning Research, 10*(Apr), 931–934.
2. Aggarwal, C., Ashish, N., & Sheth, A. (2013). The internet of things: A survey from the data-centric perspective. In *Managing and mining sensor data* (pp. 383–428). Springer, Boston, MA.
3. Al-Turjman, F., & Alturjman, S. (2018). Confidential smart-sensing framework in the IoT era. *The Journal of Supercomputing, 74*(10), 5187–5198. https://doi.org/10.1007/s11227-018-2524-1.
4. Al-Turjman, F., & Alturjman, S. (2018). Context-sensitive access in industrial internet of things (IIoT) healthcare applications. *IEEE Transactions on Industrial Informatics, 14*(6), 2736–2744. https://doi.org/10.1109/TII.2018.2808190.
5. Al-Turjman, F., Hasan, M. Z., & Al-Rizzo, H. (2018). Task scheduling in cloud-based survivability applications using swarm optimization in IOT. *Transactions on Emerging Telecommunications Technologies 0*(0), e3539. https://doi.org/10.1002/ett.3539. https://onlinelibrary.wiley.com/doi/abs/10.1002/ett.3539.
6. Alabady, S. A., & Al-Turjman, F. (2018). Low complexity parity check code for futuristic wireless networks applications. *IEEE Access, 6,* 18,398–18,407. https://doi.org/10.1109/ACCESS.2018.2818740.
7. Alabady, S. A., Al-Turjman, F., & Din, S. (2018). A novel security model for cooperative virtual networks in the IoT era. *International Journal of Parallel Programming.* https://doi.org/10.1007/s10766-018-0580-z.
8. Amaxilatis, D., Boldt, D., Choque, J., Diez, L., Gandrille, E., Kartakis S., et al. (2018). Advancing experimentation-as-a-service through urban IoT experiments. *IEEE Internet of Things Journal.* https://doi.org/10.1109/JIOT.2018.2871766.
9. Barowy, D. W., Curtsinger, C., Berger, E. D., & McGregor, A. (2016). Automan: A platform for integrating human-based and digital computation. *Communications of the ACM, 59*(6), 102–109. https://doi.org/10.1145/2927928. https://doi.org/10.1145/2927928.
10. Bello, J. P., Silva, C. T., Nov, O., DuBois, R. L., Arora, A., Salamon, J., et al. (2018). SONYC: A system for the monitoring, analysis and mitigation of urban noise pollution. *CoRR abs/1805.00889.* http://arxiv.org/abs/1805.00889.

11. Bischof, S., Karapantelakis, A., Nechifor, C., Sheth, A. P., Mileo, A., Barnaghi, P. (2014). Semantic modelling of smart city data. *Presented at the W3C Workshop on the Web of Things, Berlin.*
12. Chilton, L. B., Little, G., Edge, D., Weld, D. S., & Landay, J. A. (2013). Cascade: Crowdsourcing taxonomy creation. In *Proceedings of the SIGCHI Conference on Human Factors in Computing Systems, CHI '13* (pp. 1999–2008). New York: ACM. 10.1145/2470654.2466265, URL http://doi.acm.org/10.1145/2470654.2466265.
13. CitySDK, City service development kit, https://www.citysdk.eu/. Acccssed 15 Dec 2018.
14. CKAN, the open source data portal, http://ckan.org. Accessed 15 Dec 2018.
15. Deligiannidou, A., Amaxilatis, D., Mylonas, G., & Theodoridis, E. (2016). Knowledge co-creation in the OrganiCity: Data annotation with JAMAiCA. In *2016 IEEE 3rd World Forum on Internet of Things (WF-IOT)* (pp. 717–722). https://doi.org/10.1109/WF-IoT.2016.7845492. doi.ieeecomputersociety.org/10.1109/WF-IoT.2016.7845492
16. FIWARE Orion Context Broker, https://github.com/telefonicaid/fiware-orion. Accessed 15 Dec 2018.
17. Ganz, F., Barnaghi, P., & Carrez, F. (2016). Automated semantic knowledge acquisition from sensor data. *IEEE Systems Journal, 10*(3), 1214–1225. 10.1109/JSYST.2014.2345843.
18. Guo, B., Wang, Z., Yu, Z., Wang, Y., Yen, N.Y., Huang, R., et al. (2015). Mobile crowd sensing and computing: The review of an emerging human-powered sensing paradigm. *ACM Computing Surveys, 48*, 7:1–7:31.
19. Guo, K., Lu, Y., Gao, H., & Cao, R. (2018). Artificial intelligence-based semantic internet of things in a user-centric smart city. *Sensors, 18*(5). https://doi.org/10.3390/s18051341. http://www.mdpi.com/1424-8220/18/5/1341.
20. Hido, S., Tokui, S., & Oda, S. (2013). Jubatus: An open source platform for distributed online machine learning. In *NIPS 2013, Workshop on Big Learning: Advances in Algorithms and Data Management, Lake Tahoe, NV*
21. Karkouch, A., Mousannif, H., Moatassime, H.A., & Noel, T. (2016). Data quality in internet of things: A state-of-the-art survey. *Journal of Network and Computer Applications, 73*, 57–81. https://doi.org/10.1016/j.jnca.2016.08.002.
22. Knerr, T. (2006). Tagging ontology - Towards a common ontology for folksonomies. http://code.google.com/p/tagont/, http://tagont.googlecode.com/files/TagOntPaper.pdf.
23. Lanza, J., Sanchez, L., Gutierrez, V., Galache, J. A., Santana, J. R., Sotres, P., et al. (2016). Smart city services over a future internet platform based on internet of things and cloud: The smart parking case. *Energies, 9*, 719. https://doi.org/10.3390/en9090719.
24. Min, W., Yu, L., Yu, L., & He, S. (2018). People logistics in smart cities. *Communications of the ACM, 61*(11), 54–59. https://doi.org/10.1145/3239546.
25. Mohammadi, M., Al-Fuqaha, A., Sorour, S., & Guizani, M. (2018). Deep learning for IoT big data and streaming analytics: A survey. *IEEE Communications Surveys Tutorials, 20*(4), 2923–2960. https://doi.org/10.1109/COMST.2018.2844341.
26. NGSI v2 Specification, http://fiware.github.io/specifications/ngsiv2/stable/. Accessed 15 Dec 2018.
27. Puiu, D., Barnaghi, P., Tonjes, R., Kumper, D., Ali, M.I., Mileo, A., et al. (2016). Citypulse: Large scale data analytics framework for smart cities. *IEEE Access, 4*, 1086–1108. https://doi.org/10.1109/ACCESS.2016.2541999.
28. Puschmann, D., Barnaghi, P., & Tafazolli, R. (2018). Using LDA to uncover the underlying structures and relations in smart city data streams. *IEEE Systems Journal, 12*(2), 1755–1766. https://doi.org/10.1109/JSYST.2017.2723818.
29. Raykar, V.C., Yu, S., Zhao, L.H., Valadez, G.H., Florin, C., Bogoni, L., et al. (2010). Learning from crowds. *Journal of Machine Learning Research, 11*, 1297–1322.
30. Sánchez, L., Muñoz, L., Galache, J.A., Sotres, P., Santana, J.R., Gutiérrez, V., et al. (2014). SmartSantander: IoT experimentation over a smart city testbed. *Computer Networks, 61*, 217–238.

31. Santos, P.M., Rodrigues, J.G.P., Cruz, S.B., Lourenco, T., d'Orey, P.M., Luis, Y., et al. (2018). PortoLivingLab: An IoT-based sensing platform for smart cities. *IEEE Internet of Things Journal, 5*, 523–532.
32. Socrata: Data-Driven Innovation of Government Programs, https://socrata.com. Accessed 15 Dec 2018.
33. Sotres, P., Santana, J.R., Sánchez, L., Lanza, J., & Muñoz, L. (2017). Practical lessons from the deployment and management of a smart city internet-of-things infrastructure: The SmartSantander Testbed Case. *IEEE Access, 5*, 14,309–14,322. https://doi.org/10.1109/ACCESS.2017.2723659.
34. Sounds of New York City project, https://wp.nyu.edu/sonyc/. Accessed 15 Dec 2018.
35. Spring Boot framework, http://spring.io/projects/spring-boot. Accessed 15 Dec 2018.
36. van Kranenburg, R., Stembert, N., Moreno, M.V., Skarmeta, A.F., López, C., Elicegui, I., et al. (2014). Co-creation as the key to a public, thriving, inclusive and meaningful EU IOT. In R. Hervás, S. Lee, C. Nugent, J. Bravo (Eds.), *Ubiquitous Computing and Ambient Intelligence* (pp. 396–403). Cham: Springer.
37. Webber, J. (2012). A programmatic introduction to Neo4J. In *Proceedings of the 3rd Annual Conference on Systems, Programming, and Applications: Software for Humanity, SPLASH '12* (pp. 217–218). New York: ACM. https://doi.org/10.1145/2384716.2384777.
38. Welinder, P., Branson, S., Belongie, S., & Perona, P. (2010). The multidimensional wisdom of crowds. In *Proceedings of the 23rd International Conference on Neural Information Processing Systems, NIPS'10* (Vol. 2) (pp. 2424–2432). New York: Curran Associates Inc.

Chapter 3
Deep Reinforcement Learning Paradigm for Dense Wireless Networks in Smart Cities

Rashid Ali, Yousaf Bin Zikria, Byung-Seo Kim, and Sung Won Kim

3.1 Introduction

3.1.1 Motivation

Future dense wireless local area networks (WLANs) are attracting significant devotion from researchers and industrial communities. IEEE working groups are expected to launch an amendment to the IEEE 802.11 (WLAN) standard by the end of 2019 [1]. The upcoming amendment, covering the IEEE 802.11ax high-efficiency WLAN (HEW), will deal with ultradense and diverse user environments for smart cities, such as sports stadiums, train stations, and shopping malls. One inspiring service is the promise of astonishingly high throughput to support extensively advanced technologies for fifth generation (5G) communications and Internet of Things (IoT). HEW is anticipated to infer the various and interesting features of both the learners' environment of a HEW device as well as device behavior in order to spontaneously control the optimal media access control (MAC) layer resource allocation (MAC-RA) [2] system parameters.

In real WLANs, the devices proficiently and dynamically manage WLAN resources, such as the MAC layer carrier sense multiple access with collision avoidance (CSMA/CA) mechanism to improve users' quality of experience (QoE)

R. Ali · Y. B. Zikria · S. W. Kim (✉)
Department of Information and Communication Engineering, Yeungnam University, Gyeongsan, South Korea
e-mail: rashid@yu.ac.kr; yousafbinzikria@ynu.ac.kr; swon@yu.ac.kr

B.-S. Kim
Department of Software and Communications Engineering, Hongik University, Seoul, South Korea
e-mail: jsnbs@hongik.ac.kr

© Springer Nature Switzerland AG 2020
F. Al-Turjman (ed.), *Smart Cities Performability, Cognition, & Security*,
EAI/Springer Innovations in Communication and Computing,
https://doi.org/10.1007/978-3-030-14718-1_3

[3]. Overall device performance depends on exploitation of the instability of network heterogeneity and traffic diversity. WLAN resources are fundamentally limited due to shared channel access and wireless infrastructures, whereas WLAN services have become increasingly sophisticated and diverse, each with a wide range of QoE requirements. Thus, for the success of the prospective HEW, it is vital to investigate efficient and robust MAC-RA protocols [2].

The main motivation for the work in this chapter is to highlight the issues and challenges in MAC-RA for upcoming HEW wireless networks for IoT. The aim of the chapter is to propose machine intelligence-enabled (MI) mechanism to increase the efficiency and robustness of the MAC layer in next generation wireless networks (i.e., HEW) to be part of IoT applications like smart cities. In order to infer the diverse and interesting features of users' environments as well as users' behaviors, future HEW must spontaneously control optimal MAC-RA system parameters. Recently, the field of deep learning (DL) has been flourishing in order to enable MI capabilities in wireless communications technologies. It is believed by researchers that WLANs can optimize performance by introducing DL into MAC layer resource allocation. Deep reinforcement learning (DRL) is one DL technique that is motivated by the behaviorist sensibility and control philosophy, where a learner can achieve an objective by interacting with the environment [4]. DRL uses specific learning models, such as the Markov decision process (MDP), the partially observed MDP (POMDP), and Q-learning (QL) [5]. DRL utilizes these techniques in applications like learning an unknown wireless network environment and resource allocation in femto/small cells in heterogeneous networks (HetNets) [5].

In this chapter one of the models of DRL, Q-learning, is employed to acquire awareness about the state evaluation of the MAC-RA in wireless networks. QL-based intelligent MAC-RA optimizes the performance of wireless networks especially in dense network environments.

3.1.2   Scope of the Chapter

As discussed earlier, the main objective of this chapter is to develop theory and method for MI about how MAC-RA channel access of the WLANs evolves and for identifying optimal parameters selection in an intelligent manner. To these ends, two specific goals are identified, which will be described next in this chapter.

The first goal of this chapter is to study and quantify the issues and challenges for MAC-RA in WLANs which poses challenges for the future smart cities, more specifically in dense wireless networks. It is found that the increase in density of network devices is directly proportional to the collision in the network, and results in network performance degradation. The traditional CSMA/CA channel access mechanism is evaluated for this purpose. The binary exponential backoff (BEB) algorithm of CSMA/CA blindly handles the collision avoidance in the traditional WLANs (for detailed discussion please refer to the Sect. 3.2.3 "Problem Statement"). A practical channel observation-based scaled backoff (COSB) algorithm, that overcomes the issues of BEB, is proposed for this purpose.

The second goal is to develop a method for network device's experience inference in order to spontaneously control the optimal MAC-RA parameters of above-mentioned COSB. In this chapter, the potentials of deep reinforcement learning paradigm are studied for performance optimization of channel observation-based MAC protocols (i.e., COSB) in dense wireless network. An *intelligent* Q-learning-based resource allocation (*i*QRA) mechanism is proposed for this purpose, where Q-learning is one of the prevailing models of DRL.

3.1.3 Contributions of the Chapter

The main contributions of this chapter are presented in the original publications I–IV of the Appendix. In detail, the contributions of this chapter are as follows:

- A practical channel observation-based scaled backoff (COSB) is proposed in publications II and III. It is illustrated that a practical channel collision probability can proficiently be measured by observing the channel status that is busy or idle. The measured channel collision probability is utilized to scale the backoff contention window (CW) in CSMA/CA. The proposed COSB mechanism enhances the performance of CSMA/CA.
- A deep reinforcement-learning paradigm is proposed for performance optimization of COSB algorithm. The proposed paradigm utilizes Q-learning, one of the prevailing deep reinforcement learning models for network inference. QL algorithms are useful to optimize the performance of a device through experience gained from the interaction with the unknown environment. A target-oriented learner can be an element of a larger behavioral system, such as a HEW device in a WLAN environment seeking to maximize its performance. Because QL finds solutions through the experience of interacting with an unknown environment, it is used to optimize the CW size adjustment in the COSB mechanism. An *intelligent* QL-based resource allocation (*i*QRA) mechanism is proposed to optimize the performance of the COSB mechanism for dense HEW networks. By using network inference, *i*QRA dynamically and autonomously controls backoff parameters (i.e., the backoff stages and CW sizes) selection in COSB.

3.2 Preliminaries

WLANs are experiencing extensive growth in Internet-centric data applications. Advanced technology markets are utilizing WLANs, and deployments are rapidly flourishing in public and private areas, like shopping malls, cafes, hotels and restaurants, bus/train stations, and airports. In addition, there is the rapid increase

of WLAN-enabled electronic devices, because consumers demand that their enter-
tainment devices be Internet-enabled for IoT applications. In order to cover new
device categories and new applications, exciting new technologies are emerging for
WLANs in order to address the need for increased network capacity and coverage,
efficient energy consumption, and ease of use. Consequently, WLANs need major
improvements in both throughput and efficiency. New technologies for WLAN
applications are continuously introduced. The IEEE standard for WLANs was
initiated in 1988 as IEEE 802.4L [6], and in 1990, the designation changed to IEEE
802.11 to form a WLAN standard. This standard describes the PHY layer [6] and
MAC sublayer specifications for portable, stationary, and mobile devices within a
local area for wireless connectivity. The IEEE 802.11ac [7] standard is the currently
implemented amendment from the 802.11 standard working group (WG) promising
data rates at gigabits per second. These modern communications standards and
technologies are steadily advancing PHY layer data rates in WLANs. This capacity
growth is achieved primarily through increased channel bandwidths and advanced
PHY layer techniques, like multiple-input, multiple-output (MIMO) and multiuser
MIMO (MU-MIMO). These modern communications technologies are advancing
PHY layer data rates in WLANs, although data throughput efficiency in WLANs
may degrade rapidly as the PHY layer data rate increases. The fundamental reason
for this degradation is that the current random access-based MAC protocol allocates
the entire channel to one user as a single source due to equally distributed time
domain contention resolution. Even if senders have a small amount (or less critical)
data to send, they still need to contend for the entire channel and get an equally
distributed time opportunity for transmission. As a result, the higher the PHY layer
data rate, the lower the throughput efficiency achieved. The strategies like channel
bonding, frame aggregation and block acknowledgment, and reverse direction
forwarding enhance the high-throughput capabilities in 802.11 MAC protocol [7,
8]. IEEE 802.11 standard-based WLANs often struggle to service diverse workloads
and data types. Since the applications are categorized into different priorities by the
access layer protocol, the method how to provide enhanced and efficient resource
allocation has become an interesting and challenging topic.

3.2.1   IEEE 802.11ax High Efficiency WLAN (HEW)

The IEEE Standards Association (IEEE-SA) approved standardization activity of
the IEEE 802.11ax TG, concerned with densely deployed WLANs, in May 2014
[1]. Calling it HEW, the scope of the IEEE 802.11ax amendment is to define
modifications for both the 802.11 PHY and 802.11 MAC layers that enable at least
fourfold improvement in the average throughput per station in densely deployed
networks. It is also assumed that it shall provide backward compatibility with
existing IEEE 802.11 devices operating in the same band. Unlike previous amend-
ments, this one focuses on improving metrics that reflect the user experience. The
desired enhancements will be made to support dense environments, such as wireless

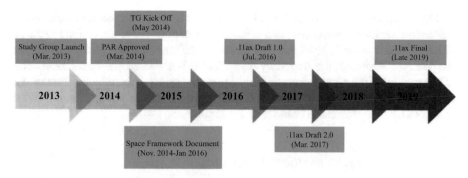

Fig. 3.1 Possible IEEE 802.11ax timeline and progress

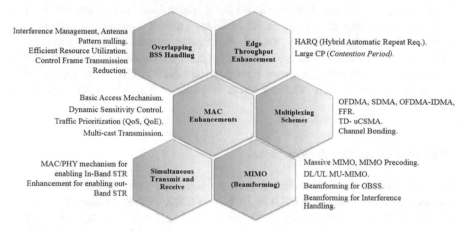

Fig. 3.2 Technologies discussed in the IEEE 802.11ax study

corporate offices, outdoor hotspots, residential apartments, and stadiums. The actual deployment of the standard is anticipated for late 2019 [9]. Figure 3.1 illustrates the possible timeline and progress towards the IEEE 802.11ax standard. The main emphasis in IEEE 802.11ax is to improve real-world performance as experienced by end users by enhancing the per-STA throughput. Possible approaches deal with three major problems in dense WLAN environments: congestion, interference, and frame conflicts. Figure 3.2 shows few of the technologies discussed in the IEEE 802.11ax TG [10].

3.2.2   MAC Layer Resource Allocation in IEEE 802.11 Wireless Networks

In wireless communication, continuous transmission is not required because devices do not use the allocated bandwidth all the time. In such cases, temporal-based access technique is favorable to IEEE 802.11 WLANs. It is observed that temporal-based resource allocation is the dominant and most important resource allocation in 802.11 WLAN. The reason is the randomness and distributed nature of the devices in the WLAN. Although the frequency-based MAC-RA and spatial-based MAC-RA allow multiuser uplink and downlink transmissions in WLANs, which is one of the promising techniques of the future WLANs, these schemes are still bound to be followed by time domain to transmit at the same time.

3.2.2.1 MAC Layer Coordination Functions

Figure 3.3 describes the MAC layer medium access coordination functions used for temporal-based resource allocation in WLANs. These functions are mainly divided into two categories: contention-based (random channel access) and contention-free (fixed assignment channel access). Both categories differ in the topological structure of the network, as well as in coordination function. These categories are further divided into distributed coordination function (DCF), enhanced DCF channel access (EDCA), point coordination function (PCF), and hybrid coordination function (HCF).

Contention-Based Coordination Functions

The initial IEEE 802.11 standard defines a contention-based distributed medium access algorithm known as DCF. The DCF uses CSMA/CA to contend for channel access. DCF can either operate under the basic access scheme (Fig. 3.4a) or the optional request-to-send/clear-to-send (RTS/CTS) (Fig. 3.4b) scheme. RTS/CTS access scheme is introduced to resolve the hidden node problem in WLANs [2]. Some device's transmissions are not detected during carrier sensing (CS) by other devices, but those transmissions interfere with transmission of other devices. These devices (STAs) are hidden from each other and can cause collisions due to unawareness of the medium access conditions. In RTS/CTS, STA transmits RTS packet and wait to receive CTS packet before actual data transmission, as shown

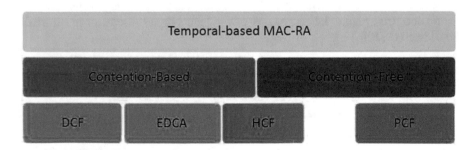

Fig. 3.3 Temporal-based MAC layer coordination functions

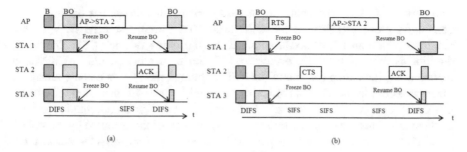

Fig. 3.4 IEEE 802.11 DCF transmission procedure: (**a**) basic access mechanism and (**b**) RTS/CTS mechanism

in Fig. 3.4b. Binary exponential backoff (BEB) and a deferral mechanism are used to differentiate the transmission start time of each device. BEB is used by STAs to contend with other STAs to access a medium and to transmit data. It is defined as the discrete backoff time slots for which the STA has to defer before accessing the wireless channel. The BEB mechanism is initiated after the channel has been found idle for a predefined distributed inter-frame space (DIFS) period. Other STAs overhear the transmission from neighboring STAs by CS and set up their network allocation vector (NAV) to avoid collisions.

In case of high traffic load and density, collisions will increase dramatically for the contending devices. EDCA is used to handle multi type of service communications in wireless networks. The main difference between DCF and EDCA is that DCF possesses only one queue for all types of data, whereas EDCA divides the coming packets into four types of logical queues, known as access categories (ACs). These ACs are defined according to the prioritized applications. The higher priority data is assigned a shorter backoff and deferral duration to obtain more chances to access the medium than other types of session. If any collision happens among more than one AC, the higher priority AC secures the opportunity to access the channel, while the others restart their backoff processes [11]. This differentiated access service for different types of application still suffers from degraded performance under densely deployed traffic situations where many contending STAs are present. That is because EDCA still follows contention-based medium access, and collisions are still possible. Moreover, high priority applications hardly provide low priority traffic any opportunity to access the resources. EDCA has this kind of unfairness [12].

Contention-Free Coordination Functions

PCF is different from the distributed medium access of DCF and EDCA, where all STAs communicate with each other via a centralized STA called a point coordinator, which usually resides in the access point (AP). The PC/AP controls the access period of the medium by splitting the resource airtime into super-frames of contention-free periods (CFPs). The controlled access period still follows the polling-based contention process (PFP). In the CFP, each STA initially sets its NAV to the maximum duration (CFPMaxDuration) at the beginning of each CFP. This NAV duration is reset if the STA receives a CF-End or a CF-End plus ACK frame from the AP, indicating the end of the CFP. Although the PCF is a contention-free medium access in the CFP, there are still several issues with providing efficiency like QoS for applications, because it does not support AC differentiation for priority applications. There is no prediction for the occupancy of the resource by the polled STA. Therefore, the fairness issue persists in the network for each STA, because the transmission time is not bounded. With densely deployed networks, this problem can cause more severe unfairness [13].

3.2.3   Problem Statement

While future physical-link technologies promise to deliver sufficient bandwidth to serve user demands, existing CSMA/CA-based channel access schemes under IEEE 802.11 are inefficient for large numbers of devices with extensively changing demands as we have seen in IoT. However, the efficiency of the current medium access protocols will soon encounter challenges when networks are deployed even more densely, like a network having to support thousands of users (STAs), or access points (APs) deployed in very close proximity to each other in smart cities, such that in a stadium, a train, or an apartment building where the density of WLAN users is very high. Most of the challenges come with the efforts to implement MAC-RA in distributed types of wireless network, specifically when there is no centralized station controlling the dedicated resource allocation and disseminating the reservation control information. One of the issues with proposing MAC-RA schemes is how to efficiently allocate available resources to the available devices. A proper MAC-RA scheme is required to serve this purpose.

The BEB scheme is the typical and traditional CSMA/CA mechanism, which was introduced in IEEE 802.11 DCF [14]. In BEB, a randomly generated backoff value for the contention procedure is used. At the first transmission attempt, the contending device generates a uniform random backoff value, from the contention window (CW). Initially, CW is set to a minimum value, and after each unsuccessful transmission, its value is doubled until it reaches the maximum defined value. Once a STA successfully transmits its data frame, CW is again reset back to the minimum value. For a network with a heavy load, resetting CW to its minimum value after successful transmission will result in more collisions and poor network performance due to an increase in probability to select similar backoff value for many STAs. Similarly, for fewer contending STAs, the blind exponential increase of CW for collision avoidance causes an unnecessarily long delay due to the wider range for selecting backoff value. Besides, this blind increase/decrease of the backoff window is more inefficient in the highly dense networks proposed for IEEE 802.11ax, because the probability of contention collision increases with the increasing number of STAs. Thus, the current MAC-RA protocol does not allow WLANs to achieve high efficiency in highly dense environments and become a part of future smart cities in IoT. Hence, to withstand this challenge, WLAN needs a more efficient and self-scrutinized backoff mechanism to promise enhanced user quality of experience (QoE).

A WLAN system performance can be severely degraded with an increase in the number of contenders, as the collision in the network is directly proportional to the density of the network. This problem statement is assured by the simulation results shown in Fig. 3.5. Figure 3.5 plots the number of STAs contending for channel access versus the average channel collision probability in a saturated (always willing to transmit) network environment with CW minimum as 32 and 64. The other simulation parameters are described in Table 3.1. The figure shows that increased network density has a direct relationship with the average channel

Fig. 3.5 Number of contending STAs vs. channel collision probability with CW minimum as 32 and 64

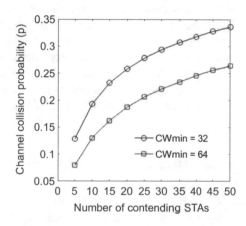

Table 3.1 MAC layer and PHY layer simulation parameters

Parameters	Values
Frequency	5 GHz
Channel bandwidth	160/20 MHz
Data rate (MCS11)	1201/54 mbps
Payload size	1472 bytes
Contention window minimum	32
Contention window maximum	1024
COSB design parameter (ω)	32
Simulation time	100/500 s
Station position	Fixed/random
Distance from AP	10/25 m
Propagation loss model	LogDistancePropagation
Mobility model	ConstantPositionMobility
Rate adaptation model	ConstantRateWifiManager MinstrelWifiManager
Error rate model	NistErrorRateModel YansErrorRateModel

collision probability; the denser the network, the higher the channel collision
probability. In such a troublesome situation, a more adaptive and self-scrutinized
MAC-RA is required by the HEW networks to maintain the performance.

3.3   Deep Reinforcement Learning Paradigm

Both academic and industrial communities have recognized that the future smart
wireless devices have to rely on enlightened learning and decision-making. Deep
learning (DL), as one of the prevailing machine learning (ML) tools, establishes an
auspicious paradigm for MAC-RA, as well. As shown in Fig. 3.6, we can imagine
an intelligent HEW device that is capable of accessing channel resources with the
aid of DL techniques. Obviously, an intelligent device learns the performance of
a specific action with the objective of preserving a specific performance metric.
Later, based on this learning, the intelligent device aims to reliably improve its
performance while executing future actions by exploiting previous experience. In
this chapter, we propose deep reinforcement learning (DRL) as a future intelligent
paradigm for MAC layer channel access in dense wireless networks for smart cities.
For this purpose, we use one of the DRL models, Q-learning, for MAC-RA in dense
wireless networks (IEEE 802.11ax HEW).

3.3.1   Deep Reinforcement Learning

DRL is motivated by behaviorist sensibility and a control philosophy, where
a learner can achieve its objective by interacting with and learning from its
surroundings. In DRL, the learner does not have clear information on whether it has
come close to its target objective. However, the learner can observe the environment

Fig. 3.6 Intelligent MAC layer resource allocation (MAC-RA) learning model

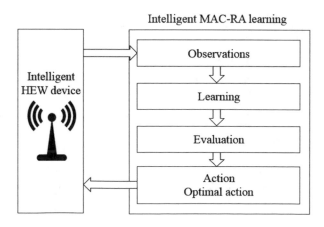

to augment the aggregate reward in an MDP [15]. DRL is one DL technique that learns about the environment, what to do, and how to outline circumstances to current actions in order to maximize a numerical reward signal. Mostly, the learner is not informed about which actions to take, yet it has to learn which actions produce the maximum reward by trying them. In the utmost exciting and inspiring situations, it is possible that actions will affect not only the instant reward but also the following state, and through that, all succeeding rewards. MDPs offer a precise framework for modeling decision making in particular circumstances, where the consequences are comparatively haphazard, and the decision maker partially governs the consequences.

Examples of DRL are partially observable MDP (POMDP) and Q-learning (QL). POMDP might be seen as speculation with MDP, where the learner is inadequate to straightforwardly perceive the original state transitions, and thus, only has constrained information. The learner has to retain the trajectory of the probability distribution of the appropriate states, based on a set of annotations, as well as the probability distribution of both the observation probabilities and the original MDP [16]. QL might be conjured up to discover an optimum strategy for taking action from any finite MDP, particularly when the environment is unknown. POMDP is an RL technique that does not follow a model, and it can also be combined with MDP models. In such a case, the QL paradigm also covers a set of states where an agent can make a decision on an action from a set of available actions. By performing an action in a particular state, the agent collects a reward, with the objective being to exploit its collective rewards. A collective reward is illustrated as a Q-function, and is updated in an iterative approach after the agent carries out an action and attains the subsequent reward [16].

The uses of POMDP paradigms create vital tools for supportive decision making in IoT systems, where the IoT devices may be considered learners, and the wireless network constitutes the environment. In a POMDP problem, the technique first postulates the environment's state space and the learner's action space, as well as endorses the Markov property among the states. Secondly, it constructs the state transition probabilities formulated as the probability of navigating from one state to another under a specific action. The third and final step is to enumerate both the learner's instant reward and its long-term reward via Bellman's equation [17]. Later, a wisely constructed iterative algorithm may be considered to classify the optimum action in each state. The applications of POMDP comprise the network selection problems of heterogeneous networks, channel sensing, and user access in CRNs. In [18], the authors proposed a mechanism for transmission power control problems of energy-harvesting systems, which were scrutinized with the help of the POMDP model. In their proposed investigation, the battery state, the channel state, and data transmission and data reception states are defined as the state space, and an action by the agent is related to transmitting a packet at a certain transmission power. QL, usually in aggregation with the MDP models, has also been used in applications of heterogeneous networks. Alnwaimi et al. presented a heterogeneous, fully distributed, multi-objective strategy for optimization of femtocells based on a QL model [19]. Their proposed model solves both the channel resource allocation and interference coordination issues in the downlink of heterogeneous femtocell networks. Their proposed model acquires channel distribution awareness, and classifies the accessibility of vacant radio channel slots for the establishment of opportunistic access. Later, it helps to pick sub-channels from the vacant spectrum pool.

3.3.2 Q-Learning as a MAC-RA Paradigm

As described in the previous section, QL has already been extensively applied in heterogeneous wireless networks. In this section, the main functional structure of the QL algorithm is described, and in a later subsection, the use of QL as a future paradigm for the backoff mechanism in a DCF is suggested for dense wireless networks.

3.3.2.1 Q-Learning Algorithm

The QL algorithm utilizes a form of DRL to solve MDPs without possessing complete information. Aside from the learner and the environment, a QL system has four main sub-elements: a policy, a reward, a Q-value function, and sometimes a model of the environment as an optional entity (most of the QL algorithms are model-free), as shown in Fig. 3.7.

Policy

The learner's way of behaving at a particular time is defined as a policy. It resembles what in psychology would be called a set of stimulus–response associations. In some circumstances, the policy can be a modest utility or a lookup table; in others, however, it may comprise extensive computations, like an exploration process. The policy is fundamental for a QL learner, in the sense that it alone is adequate to determine behavior. Generally, policies might be stochastic. A policy decides which action to take in which state.

Fig. 3.7 Q-learning model environment for an intelligent HEW

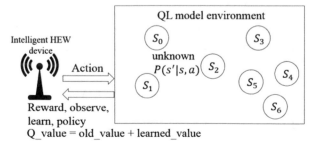

Reward

A reward expresses the objective in a QL problem. In each time step, the environment passes to the QL learner a particular quantity called the reward. The learner's exclusive goal is to exploit the total reward it obtains over the long run. The reward describes the pleasant and unpleasant events for the learner. Reward signals are the instant and crucial topographies of the problem faced by the learner. The reward signal is the key basis for changing the policy. For example, if the current action taken by a policy is followed by a low reward, a learner may decide to select other actions in future.

Q-Value Function

Though the reward specifies what is good at one instant, a value function stipulates what is good in the end. Thus, the value (known as the Q-value) of a state is the aggregate amount of rewards a learner can presume to accumulate over the future, starting from that state. For example, a state might continuously produce a low instant reward, but still have a high Q-value because it is repeatedly trailed by other states that produce high rewards. To make a WLAN environment correspondent, rewards are somewhat like a high channel collision probability (unpleased) and a low channel collision probability (pleased), whereas Q-values resemble a more sophisticated and prophetic verdict of how pleased or unpleased the learner (STA) is in a particular state (e.g., the backoff stage). If there is no reward, there will be no Q-value, and the only purpose for estimating the Q-value is to attain additional rewards. It is the Q-value that a learner is the most anxious about when making and assessing verdicts. A learner selects optimum actions based on Q-value findings. It seeks actions that carry states of a maximum Q-value, not a maximum reward, because these actions attain the highest amount from the rewards for the learner over the long run.

Environment Model

An optional element of QL is a model of the system, which somewhat mimics the performance of the environment. Typically, it allows inferences to be made about how the environment will perform. For example, given a state and an action, the model might envision the subsequent state and the next reward. Environment models are used for planning a way to decide on a sequence of actions by considering latent future situations. In an example of a WLAN system, a device (the learner) would like to plan its future decisions based on a given state (e.g., the backoff stage) and action, along with its rewards (e.g., channel collision probability).

3.3.2.2   Scope and Limitations of QL

As discussed above, QL depends strongly on the notion of the state as input to the policy and the Q-value function. Informally, we can think of the state as a flag passing to the learner with some sense of how the environment is at a specific time. A large portion of QL techniques are organized around evaluating Q-value functions; however, it is not entirely essential to do this to take care of DRL problems. For instance, approaches like genetic algorithms, genetic programming, simulated forging, and other optimization algorithms have been utilized to approach DRL problems while never engaging value functions [20]. These evolutionary approaches assess the lifetime conduct of numerous non-learners, each utilizing an alternate policy for interfacing with the environment and selecting those actions that are able to acquire the most rewards. If the space of policies is adequately small, or can be organized so that the best policies are common or simple to discover, or if a considerable measure of time is available for the search, then evolutionary approaches can be viable. Furthermore, evolutionary approaches have focal points for problems in which the learner cannot detect the entire state of the environment. In contrast to evolutionary approaches, QL techniques learn while interfering with the environment. Techniques ready to exploit the details of individual behavioral interactions can be substantially more productive than evolutionary strategies in many types of wireless network.

3.4   *Intelligent* Q-Learning-Based Resource Allocation (*i*QRA)

The QL-based MAC-RA scheme can be used to guide future densely deployed WLAN devices and to allocate channel resources more efficiently. When a WLAN device is deployed in a new environment, usually no data are available on historical scenarios. Therefore, QL algorithms are the best choice to observe and learn the environment for optimal policy selection. In a densely deployed WLAN, channel collision is the most vital issue causing performance degradation. Since QL finds solutions through the experience of interacting and learning with an environment, it is proposed using it to model the optimal contention window in MAC-RA. In other words, a learner (STA) controls the CW selection intelligently with the aid of the QL-based algorithm. In a DCF-based backoff mechanism, STAs can be equipped with the intelligent QL algorithm. The policy is the decision of an STA to change the CW size (i.e., to take an action to move to the next backoff stage and increase the CW size, or to move to the previous/first backoff stage and decrease/reset the CW size, or possibly stay at the same backoff stage with no change to the CW). The reward function captures the gain of each action performed in any state. For example, a reward can be channel collision probability, channel access delay, or the packet loss ratio (PLR) experienced by the STA at a specific state (backoff stage).

Next subsection elaborates a channel observation-based scaled backoff (COSB) protocol to tackle the blindness problem of conventional BEB algorithm. Later, in the second subsection, an *intelligent* QL-based resource allocation (*i*QRA) is proposed to optimize the performance of COSB.

3.4.1 Channel Observation-Based Scaled Backoff (COSB) Mechanism

In the proposed COSB protocol, after the communication medium has been idle for a DIFS period, all the STAs competing for a channel proceed to the backoff procedure by selecting a random backoff value B as shown in Fig. 3.8. The time immediately following an idle DIFS is slotted into observation time slots (ξ). The duration of ξ is either a constant slot time σ during an idle period or a variable busy (successful or collided transmission) period. While the channel is sensed to be idle during σ, B decrements by one. A data frame is transmitted after B reaches zero. In addition, if the medium is sensed to be busy, the STA freezes B and continues sensing the channel. If the channel is again sensed to be idle for DIFS, B is resumed. Each individual STA can proficiently measure channel observation-based conditional collision probability p_{obs}, which is defined as the probability that a data

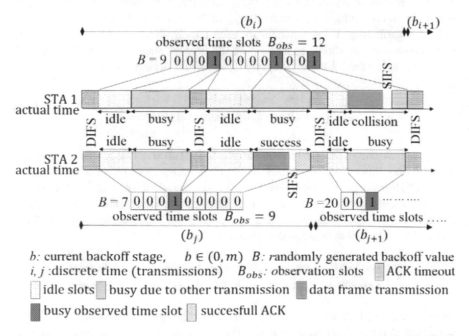

Fig. 3.8 Channel observation mechanism of channel observation-based scaled backoff (COSB) during the backoff procedure

frame transmitted by a tagged STA fails. The time is discretized in B_{obs} observation time slots, where the value of B_{obs} is the total number of ξ observation slots between two consecutive backoff stages as shown in Fig. 3.8. A tagged STA updates p_{obs} from B_{obs} of backoff stage b_i at the ith transmission as,

$$p_{obs} = \frac{1}{B_{obs}} \times \sum_{k=0}^{B_{obs}-1} S_k, \tag{3.1}$$

where for an observation time slot k, $S_k = 0$ if ξ is empty (idle) or the tagged STA transmits successfully, while $S_k = 1$ if ξ is busy or the tagged STA experiences collision as shown in Fig. 3.8. In the figure, STA 1 randomly selects its backoff value $B = 9$ for its b_i backoff stage. Since STA 1 observes nine idle slot times, two busy periods, and one collision ($B_{obs} = 9 + 2 + 1 = 12$), p_{obs} is updated as $\frac{2+1}{B_{obs}} = \frac{3}{12} = 0.25$ in the next backoff stage b_{i+1}.

According to the channel observation-based conditional collision probability p_{obs}, the adaptively scaled contention window value is $W_{b_{i+1}}$ at backoff stage b_{i+1} of the transmission time $i + 1$, where $b_{i+1} \in (0, m)$ for the maximum m number of backoff stages, and i is the discretized time for the data frame transmissions of a tagged STA. More specifically, when a transmitted data frame has collided, the current contention window W_{b_i} of backoff stage b_i at the ith transmission time slot is scaled-up according to the observed p_{obs} at the ith transmission, and when a data frame is transmitted successfully, the current contention window W_{b_i} is scaled-down according to the observed p_{obs} at the ith transmission. Unlike the BEB (where backoff stage is incremented for each retransmission and resets to zero for new transmission as shown in Fig. 3.9a), the backoff stage b_i in COSB at the ith transmission has the following property of increment or decrement:

$$b_{i+1} = \begin{cases} \min[b_i + 1, m], \text{ collision at } i\text{th transmit} \\ \max[b_i - 1, 0], \text{ success at } i\text{th transmit} \end{cases}. \tag{3.2}$$

Figure 3.9b shows that the backoff stage in COSB does not reset after a successful transmission. Since the current backoff stage represents the number of collisions or successful transmissions of a tagged STA, it helps to scale the size of CW efficiently. The incremented or decremented backoff stage b_i results in scaling-up or scaling-down of the current contention window, respectively. The scaling-up and scaling-down of the contention window operates as follows:

$$W_{b_{i+1}} = \begin{cases} \min\left[2^{b_{i+1}} \times W_{min} \times \omega^{p_{obs}}, W_{max}\right], \text{ collision at } i\text{th transmit} \\ \max\left[2^{b_{i+1}} \times W_{min} \times \omega^{p_{obs}}, W_{min}\right], \text{ success at } i\text{th transmit} \end{cases}, \tag{3.3}$$

where ω is a constant design parameter to control the optimal size of the contention window and is expressed as $\omega = W_{min}$. The W_{min}, and W_{max} are the minimum CW and maximum CW values.

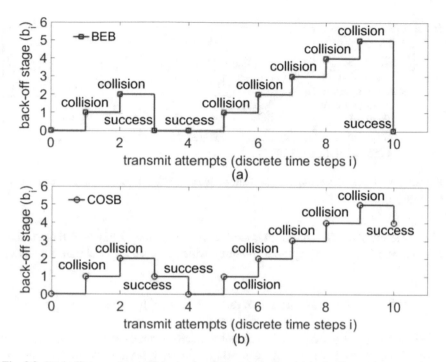

Fig. 3.9 Backoff stage after collision/successful transmission; (**a**) backoff stage increment/reset in binary exponential backoff (BEB); and (**b**) backoff stage increment/decrement in COSB

3.4.2 iQRA Algorithm

The proposed *i*QRA mechanism in COSB, consists of a set of states S (backoff stages), where an intelligent HEW device performs an action a (increments/decrements) according to COSB mechanism. By performing action a following a policy Φ in a particular state s, the device collects a reward r, that is $r(s, a)$ with the objective to exploit the collective reward, $Q(s, a)$ which is a Q-value function. Figure 3.10 depicts the model environment with its elements for the proposed *i*QRA mechanism.

Let $S = \{0, 1, 2, \ldots, m\}$ denotes a finite set of m possible states of the HEW environment, and let $A = \{0, 1\}$ represents a finite set of permissible actions to be taken, where zero indicates decrement, and 1 indicates increment (As described earlier, in COSB, there are two possible actions: increase or decrease of the backoff stage). At time slot t, the STA observes the current state (s), that is, $s_t = s \in S$, and takes an action (a), that is, $a_t = a \in A$ based on policy Φ. As mentioned before, the default policy of a device in COSB is to increment its state if collision happened and decrement for successful transmission. Thus, action a_t changes the environmental state from s_t to $s_{t+1} = s' \in S$ according to $\Phi(a|s) = \begin{cases} s' = s + 1, \text{if collision} \\ s' = s - 1, \text{if succesful} \end{cases}$.

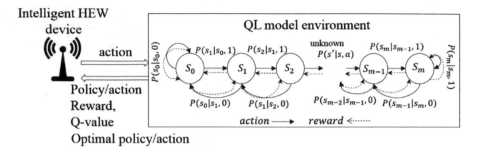

Fig. 3.10 *Intelligent* Q-learning-based resource allocation (*i*QRA); the system environment and its elements

The objective of the QL algorithm is to discover an optimal policy Φ^{opt} that exploits the total expected reward (optimal Q-value), which is given by a Bellman's equation [17]:

$$Q^{\text{opt}}(s, a) = \mathbb{E}\left\{ r_t(s, a) + \beta \times \max_{a'} Q^{\text{opt}}(s', a') \mid s_t = s, a_t = a \right\} \quad (3.4)$$

Since the reward may easily get unbounded, a discounted reward factor, β ($0 < \beta < 1$), is used. In the QL algorithm, $Q(s, a)$ estimates the reward as the cumulative reward and is updated as follows:

$$Q(s, a) = (1 - \alpha) \times Q(s, a) + \alpha \times \Delta Q(s, a) \quad (3.5)$$

where α is the learning rate and is defined as $0 < \alpha < 1$. The learning occurs quickly, based on the improved learning estimate, $\Delta Q(s, a)$, and is expressed as

$$\Delta Q(s, a) = \left\{ r(s, a) + \beta \times \max_a Q(s', a') \right\} - Q(s, a) \quad (3.6)$$

The $\max_a Q(s', a')$ defines the best-estimated value for the prospective state, s'. In the long run, $Q(s, a)$ converges to the optimal Q-value $Q^{\text{opt}}(s, a)$, that is, $\lim_{t \to \infty} Q(s, a) = Q^{\text{opt}}(s, a)$. The naivest policy for action selection can be to pick one of the actions with the maximum measured Q-value (i.e., exploitation). This exploitation method follows the optimal policy Φ^{opt} and is known as a greedy action ($a^{\Phi^{\text{opt}}}$) selection method which can be written as

$$a^{\Phi^{\text{opt}}} = \Phi^{\text{opt}}(a|s) = \text{argmax}_a \, Q^{\text{opt}}(s, a) \quad (3.7)$$

where argmax_a signifies $a^{\Phi^{\text{opt}}}$, for which the expression that follows is exploited. The immediate reward is maximized by continuous exploitation of the greedy action selection method. A simple substitute is to exploit most of the time, but every once in a while, the STA explores all the permissible actions independent of $a^{\Phi^{opt}}$, that is with default policy Φ (known as exploration) with probability ε. The greedy and non-greedy selection of actions is known as the ε-greedy method [17]. The main feature of the ε-greedy technique is that, as the number of instances increases, every action guarantees the convergence of learning estimate $gQ(s, a)$. In HEW, for dense WLANs, the STA would exploit to improve throughput performance, and would explore to know the dynamicity of the WLAN environment.

COSB conducts p_{obs} at every transmission attempt (state), which can be considered as reward of the action. Therefore, the p_{obs}, from Eq. (3.1) is expressed as the reward in order to minimize channel collision probability. The reward given by action a_t taken in state s_t at time t is described as

$$r_t(s_t, a_t) = 1 - p_{obs} \qquad (3.8)$$

The above statement indicates how pleased the STA was with its action in state s_t. In Fig. 3.10, the STA moves from one state to another state with $1 - p_{obs}$ as a reward. The STA observes and learns the environment to optimize the backoff process. Algorithm 1 depicts the steps performed by the proposed iQRA mechanism to optimize the COSB protocol.

Algorithm 1 COSB performance optimization using iQRA

1: **GLOBAL:** Initialize $r(s, a)$, and $Q(s, a)$.
2: **Function:** Select CW using iQRA
   Input: p_{obs}
   Output: Optimized CW
3:   **Initialize:** cur_rew $= 0$, $\Delta Q(s, a) = 0$
4:   Calculate reward according to equation (4.31)
5:   Update reward matrix for $r(s, a) = $ cur_rew
6:   Calculate improved learning estimate $\Delta Q(s, a)$ according to equation (4.29)
7:   Update $Q(s, a)$ according to equation (4.28)
8:   Pick a random value to explore or exploit (ε-greedy method)
9:   **if** (exploit)
10:     Find $a^{\Phi^{opt}}$ according to equation (4.30)
11:     Scale CW according to the optimal action
12:   **else** (explore)
13:     Scale CW using COSB mechanism
14:   **end if**
15:   **return** CW
16: **end Function**

3.5 Performance Evaluation

3.5.1 Simulation Scenarios and Parameters

The proposed learning-based iQRA mechanism is simulated using the ns-3 network simulator, version 3.28 [21], with IEEE 802.11ax HEW indoor scenario model for dense WLANs (Typically suitable for office-buildings in smart cities). Some important simulation parameters are given in Table 3.1.

To evaluate the QL parameter selection for the proposed iQRA, 25 contending STAs are simulated for 100 s, varying α and β with small (0.2), medium (0.5) and large (0.8) values. Probability ε was set to 0.5 for balanced exploration and exploitation. Figure 3.11 shows the convergence of learning estimate ΔQ from Eq. (3.6) with respect to the learning rate (α). The figure depicts how a larger α makes ΔQ converge faster. The convergence of ΔQ indicates that the STA has learned its environment and can exploit optimal actions in the future. An interesting observation is that ΔQ is not steady in the beginning, which is due to the initial exploration of the environment. Therefore, most of the states do not optimize the value function in the beginning. Later, the STA locates the states that can deliver the most rewards, increasing the cumulative reward. After enough instances (13 for $\alpha = 0.8$ in Fig. 3.11), it can be seen that the learner has found configurations that can lead to optimization of the process. Similarly, we observe in Fig. 3.12 that ΔQ converges faster with the small value for β, compared to the larger values. In both cases (Figs. 3.11 and 3.12), ε was set to 0.5, indicating equal opportunities for exploration and exploitation. The large value for α and the small value for β (along with equal probability ε) yield the best results for optimization in the system. The convergence of learning estimates shows that an optimal solution for the environment exists.

Figures 3.13 and 3.14 portrays the effects of the parameters on throughput of the system (Fig. 3.13 for a small network of 15 STAs, and Fig. 3.14 for a dense network

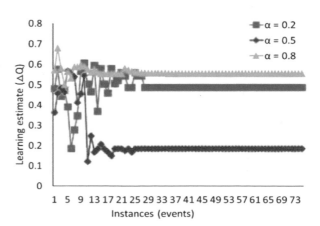

Fig. 3.11 Convergence graphs of the learning estimate (ΔQ) from varying the learning rate, α ($\beta = 0.2$)

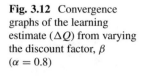

Fig. 3.12 Convergence graphs of the learning estimate (ΔQ) from varying the discount factor, β ($\alpha = 0.8$)

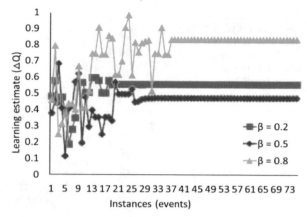

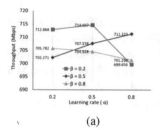

(a)   (b)   (c)

Fig. 3.13 Throughput comparison of α and β in a small network of 15 STAs with (**a**) $\varepsilon = 0.2$, (**b**) $\varepsilon = 0.5$, and (**c**) $\varepsilon = 0.8$

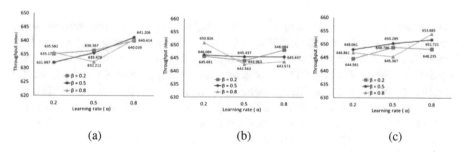

(a)   (b)   (c)

Fig. 3.14 Throughput comparison of α and β in a dense network of 50 STAs with (**a**) $\varepsilon = 0.2$, (**b**) $\varepsilon = 0.5$, and (**c**) $\varepsilon = 0.8$

of 50 STAs). As shown in Fig. 3.13a, if ε is set to 0.2 for a small network of 15 STAs, $\alpha = 0.5$ and $\beta = 0.2$ give the best results. However, in this case, decreasing α (i.e., $\alpha = 0.2$) has little effect on throughput, but increasing it to $\alpha = 0.8$ degrades throughput dramatically. Figure 3.13b shows that if ε and α are set to 0.5, β can be set small, medium, or large. However, for $\varepsilon = 0.8$ and $\alpha = 0.5$, seting β to its medium value (i.e., $\beta = 0.5$) enhances throughput, as shown in Fig. 3.13c. Figures 3.14a–c show that for a dense network system of 50 STAs, a small value

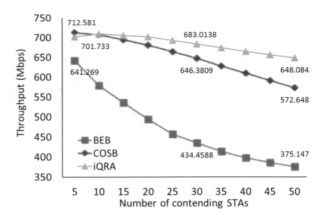

Fig. 3.15 Throughput comparison of BEB, COSB, and *i*QRA with $\alpha = 0.2$, $\beta = 0.8$ and $\varepsilon = 0.5$ in a network of five to 50 contending STAs

for α (i.e., $\alpha = 0.2$) and a large value for β (i.e., $\beta = 0.8$) are efficient for small and medium values of ε (i.e., $\varepsilon = 0.2$ and $\varepsilon = 0.5$). With a large value for ε (i.e., $\varepsilon = 0.8$), as shown in Fig. 3.14c, throughput is improved if the large α and β are used (i.e., $\alpha = 0.8$ and $\beta = 0.8$). Thus, from Figs. 3.13 and 3.14, we show that a combination of a small α, a large β, and a medium value for ε (i.e., $\alpha = 0.82$ $\beta = 0.8$, and $\varepsilon = 0.5$) is somewhat efficient for both sparse and dense network systems.

3.5.2 Throughput

To evaluate the performance of *i*QRA, simulation results are compared with the state-of-the-art BEB and non-intelligent COSB algorithms. Figure 3.15 shows how the *i*QRA mechanism optimizes the throughput of COSB, specifically in a dense network of 50 contending STAs. The performance improvement clearly indicates that the QL-based proposed mechanism is effective at learning the wireless network. In a network of five contending STAs, *i*QRA achieves relatively lower system throughput than COSB. COSB outperforms for very small networks (i.e., for less than ten contending STAs). The performance of *i*QRA may degrades in small networks due to low and irregular rewards.

3.5.3 Average Channel Access Delay

The channel access delay for a successfully transmitted data frame is defined as the interval from the time the frame is at the head of the queue (ready for transmission) until successful acknowledgement that the frame was received. If a frame reaches the given retry limit, it is dropped, and its time delay is not included in the calculation of channel access delay. Figure 3.16 depicts the performance of

Fig. 3.16 Channel access delay comparison of BEB, COSB, and iQRA with $\alpha = 0.2$, $\beta = 0.8$, and $\varepsilon = 0.5$ in a network of five to 50 contending STAs

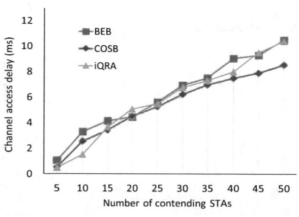

Fig. 3.17 The number of successfully transmitted packets by each STA in a dense network of 50 STAs

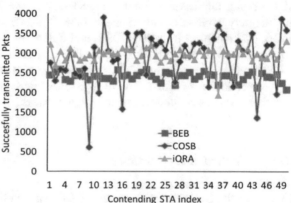

the proposed iQRA mechanism along with the conventional BEB and the original COSB mechanisms in terms of channel access delay (in milliseconds). From the figure, it can be observed that the proposed iQRA mechanism has a higher channel access delay, compared to COSB; however, it does not exceed the conventional BEB mechanism. It is obvious that the iQRA mechanism has an increased channel access delay due to its environment inference characteristics.

3.5.4 Fairness

The fairness issue can be seen for COSB in Fig. 3.17. In a dense network environment of 50 STAs, COSB suffers from the fairness problem due to some STAs continuously operating at a higher CW size, and a few fortunate STAs can operate at a lower CW size. Under COSB, once the STA reaches a larger CW, it has to transmit successfully many times to return to the smaller CW, which seems difficult in a dense network environment due to the high probability of collision. The proposed

Table 3.2 Jain's fairness index (JFI) comparison

STAs	BEB	COSB	iQRA
5	0.999	0.953	0.999
10	0.999	0.999	0.999
15	0.999	0.999	0.999
20	0.999	0.997	0.999
25	0.999	0.992	0.999
30	0.998	0.993	0.999
35	0.998	0.991	0.999
40	0.997	0.990	0.999
45	0.998	0.948	0.999
50	0.998	0.953	0.998

iQRA brings fairness to the contending STAs, because every STA autonomously and intelligently exploits its environment. Table 3.2 shows the values in Jain's fairness index [22] achieved by BEB, COSB, and iQRA for a small network of five STAs to a large, dense network of 50 STAs. It is observed that the previously proposed COSB mechanism was unfair for small to large network environments, while the iQRA mechanism optimizes COSB to perform fairly among the contending STAs, whether it is for a small network or a large network.

3.5.5 Network Dynamicity

Subsequently, QL is essentially intended to make intelligent adjustments according to the dynamics of the environment. A dynamic environment can be the activation of more contending STAs in the network or the deactivation of previously active STAs. This thesis evaluates the performance of the proposed iQRA mechanism by activating five more contending STAs every 50 s until the number of STAs reached 50. Figure 3.18 explains the effects of network dynamics on ΔQ (i.e., learning estimate) of a tagged STA. The figure shows 1400 learning instances (events) of a tagged STA during the simulation period (500 s). Each instance represents the updated value of learning estimate ΔQ whenever a packet transmission is attempted. As shown in the figure, with changes in the number of contending STAs within the network, the tagged STA experiences a fluctuation in ΔQ, indicating the change in the environment. Later, this QL-equipped, intelligent tagged STA converges and is capable of optimizing the performance in a dynamic wireless environment. In Fig. 3.19, it can be seen that iQRA eventually reaches a steady state in system throughput. On the other hand, BEB and COSB are severely affected by the increase in the number of competing STAs.

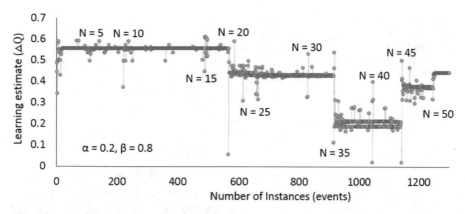

Fig. 3.18 Convergence of the learning estimates (ΔQ) in a dynamic network environment (increasing the number of contenders every 50 s)

Fig. 3.19 System throughput comparison in a dynamic network environment (increasing contenders by five every 50 s)

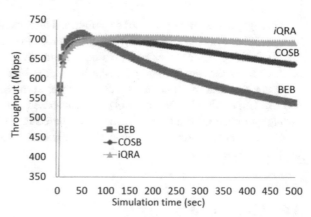

3.5.6   Distance-Based Rate Adaptation Models

Throughput shown in Figs. 3.15 and 3.19 are achieved in a network environment using the ConstantRateWifiManager rate-adaptation algorithm [21], in which contending STAs are placed at a fixed distance from the access point (AP). Hence, all the devices are transmitting at a constant data rate. To evaluate the performance of the proposed iQRA algorithm, we simulated a more practical and real network environment, such as MinstrelWifiManager [21]. The Minstrel rate adaptation varies the transmission rate of the sender STA to match the WLAN channel conditions (mainly based on the distance from the AP), in order to achieve the best possible performance. The results shown in Fig. 3.20 are achieved in an IEEE 802.11a (54 Mbps) wireless network for $N = 10$. All contending STAs were randomly placed within a distance of 25 m from the AP. A tagged STA (initially placed at a 1 m distance) moves away from the AP.

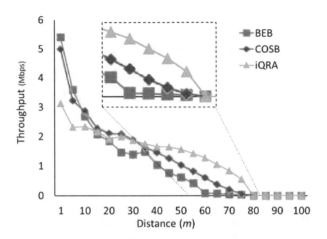

Fig. 3.20 Throughput comparison for distance-based rate-adaptation network environment

Throughput shown in Fig. 3.20 was obtained after each 5 m distance from the AP. The performance of a tagged STA for all three of the compared algorithms (BEB, COSB, and *i*QRA) degrades as the distance from the AP increases, as shown in Fig. 3.20. Observe that the throughput of the STA for BEB is close to zero after the STA reaches a distance of 60 m, and finally becomes zero when it exceeds the coverage (80 m). Under COSB, due to its observation-based nature, a STA achieves higher throughput even after a 60 m distance, compared to BEB. However, the proposed *i*QRA maintains performance, even if the distance increases to 80 m, due to its intelligence capability.

3.6 Conclusion

The upcoming dense high-efficiency WLAN (i.e., IEEE 802.11ax HEW) promises per-device throughput performance that is four times higher for 5G and IoT technologies. One of the bottlenecks for this performance achievement is tackling the huge challenge of efficient MAC layer resource allocation in WLANs due to their distributed contention-based nature. Currently, the CSMA/CA-based WLAN uses a binary exponential backoff mechanism (BEB), which blindly increases and decreases the contention window after collisions and successful transmissions, respectively. To handle the performance degradation challenge caused by the increasing density of WLANs, a self-scrutinized channel observation-based scaled backoff (COSB) mechanism based on a practical channel collision probability is proposed in this dissertation. COSB overcomes the limitation of BEB to achieve high efficiency and robustness in highly dense networks, and enhances the performance of CSMA/CA in dense networks. However, to satisfy the diverse requirements of such dense WLANs, it is anticipated that prospective WLANs will autonomously access the best channel resources with the assistance of sophisticated wireless channel condition inference. Motivated by the potential applications and features of deep reinforcement learning (DRL) in wireless networks, such as the deployment of cognitive radio. In this chapter, one of the DRL techniques, Q-learning, is proposed as an intelligent paradigm for MAC layer resource allocation in dense WLANs. The proposed DRL paradigm uses intelligent QL-based inference to optimize the performance of COSB, named as *intelligent* QL-based resource allocation (*i*QRA). Simulation results show that the proposed *i*QRA mechanism optimizes the performance of COSB in fixed wireless STA network environments, as well as for randomly placed and distance-based rate adaptation network environments.

Appendix: SCI/SCIE Journal Publications Related to the Chapter

1. R. Ali, S. W. Kim, B. Kim, and Y. Park, "Design of MAC layer resource allocation schemes for IEEE 802.11ax: Future directions," *IETE Technical Review*, vol. 35(1), pp. 28–52, February 2018.
2. R. Ali, N. Shahin, R. Bajracharya, B. Kim and S. W. Kim, "A self-scrutinized backoff mechanism for IEEE 802.11ax in 5G unlicensed networks," *Sustainability*, vol. 10(4), pp. 1–15, April 2018.
3. R. Ali, N. Shahin, Y. Kim, B. Kim and S. W. Kim, "Channel observation-based scaled backoff mechanism for high-efficiency WLANs," *Electronics Letters*, vol. 54(10), pp. 663–665, May 2018.
4. R. Ali, N. Shahin, Y. B. Zikria, B. Kim, and S. W. Kim, "Deep reinforcement learning paradigm for performance optimization of channel observation-based MAC protocols in dense WLANs," *IEEE Access*, (accepted).

References

1. IEEE 802.11ax High Efficiency WLAN (HEW), P802.11-TGax, 2014.
2. Ali, R., Kim, S. W., Kim, B., & Park, Y. (2018, February). Design of MAC layer resource allocation schemes for IEEE 802.11ax: Future directions. *IETE Technical Review, 35*(1), 28–52. https://doi.org/10.1080/02564602.2016.1242387
3. Ali, R., Shahin, N., Bajracharya, R., Kim, Y. T., Kim, B., & Kim, S. W. (2018). A self-scrutinized backoff mechanism for IEEE 802.11ax in 5G unlicensed networks. *Sustainability, 10*(4), 1201. https://doi.org/10.3390/su10041201
4. Moon, J., & Lim Y (2017). A reinforcement learning approach to access management in wireless cellular networks. *Wireless Communications and Mobile Computing, 2017*, Article ID 6474768, 7 pages. https://doi.org/10.1155/2017/6474768
5. Jiang, C., Zhang, H., Ren, Y., Han, Z., Chen, K., & Hanzo, L. (2017, April). Machine learning paradigms for next-generation wireless networks. *IEEE Wireless Communications, 24*(2), 98–105. https://doi.org/10.1109/MWC.2016.1500356WC
6. Primer. (2013). *Wi-Fi: overview of the 802.11 physical layer and transmitter measurements* (pp. 4–7). Beaverton: Tektronix Inc.

7. Gong, M. X., Hart, B., & Mao, S. (2015, January). Advanced wireless LAN Technologies: IEEE 802.11ac and beyond. *GetMobile: Mobile Computing and Communications, 18*(4), 48–52.
8. Charfi, E., Chaariand, L., & Kamoun, L. (2013, November). PHY/MAC enhancements and QoS mechanisms for very high throughput WLANs: A survey. *IEEE Communications Surveys and Tutorials, 15*(4), 1714–1735.
9. Barber, P. (2013). *IEEE 802.11ax project plan*, IEEE technical presentation [Online]. Available: http://www.ieee802.org/11/Reports/tgax_update.htm
10. Yunoki, K., & Misawa, Y. (2013). *Possible approaches for HEW*, IEEE technical presentation [Online]. Available: http://www.ieee802.org/11/Reports/tgax_update.htm
11. IEEE MAC Enhancement for Quality of Service. IEEE Standard 802.11e, 2005.
12. Yu, X., Navaratnam, P., & Moessner, K. (2013, July). Resource reservation schemes for IEEE 802.11-based wireless networks: A survey. *IEEE Communications Surveys and Tutorials, 15*(3), 1042–1061.
13. Youssef, M. A., & Miller, R. E. (2002). *Analyzing the point coordination function of the IEEE 802.11 WLAN protocol using a systems of communicating machines specification, UMIACS Technical Report CS-TR-4357* (p. 36). College Park, MD: UM Computer Science Dept.
14. IEEE Standard for Information Technology. Part 11: Wireless LAN medium access control (MAC) and physical layer (PHY) specifications. ANSI/IEEE Std 802.11 2007, i-513.
15. Li, R., Zhao, Z., Zhou, X., Ding, G., Chen, Y., Wang, Z., et al. (2017). Intelligent 5G: When cellular networks meet artificial intelligence. *IEEE Wireless Communications, 24*(5), 175–183. https://doi.org/10.1109/MWC.2017.1600304WC
16. Alpaydin, E. (2014). *Introduction to machine learning* (3rd ed.). Cambridge, MA: MIT Press. ISBN: 978-0-262-028189.
17. Sutton, R. S., & Barto, A. G. (1998). *Reinforcement learning: An introduction* (2nd ed.). Cambridge, MA: MIT Press. ISBN: 0262193981.
18. Aprem, A., Murthy, C. R., & Mehta, N. B. (2013). Transmit power control policies for energy harvesting sensors with retransmissions. *IEEE Journal of Selected Topics in Signal Processing, 7*(5), 895–906. https://doi.org/10.1109/JSTSP.2013.2258656
19. Alnwaimi, G., Vahid, S., & Moessner, K. (2015). Dynamic heterogeneous learning games for opportunistic access in LTE-based macro/femtocell deployments. *IEEE Transactions on Wireless Communications, 14*(4), 2294–2308. https://doi.org/10.1109/TWC.2014.2384510
20. Ali, R., Shahin, N., Kim, Y. T., Kim, B., & Kim, S. W. (2018, May). Channel observation-based scaled backoff mechanism for high efficiency WLANs. *Electronics Letters, 54*(10), 663–665.
21. The Network Simulator — ns-3 [Online]. Available: https://www.nsnam.org/
22. Jain, R., Chiu, D., & Hawe, W. (1984). *A quantitative measure of fairness and discrimination for resource allocation in shared computer system*. Hudson: Eastern Research Laboratory, Digital Equipment Corporation.

Chapter 4
Energy Demand Forecasting Using Deep Learning

Bahrudin Hrnjica and Ali Danandeh Mehr

Introduction to Machine Learning

Systems of intelligent behavior have been the subject of interest for scientists over the last few decades. They have tried to integrate intelligence through adaption, learning, autonomy, and solving complex problems. Such research led to the emergence of a new scientific field that is now called artificial intelligence (AI) [28]. AI can be described as an action performed by a machine that can be characterized as intelligent, since if a human had to apply the same action, intelligence must be used to achieve the same goal. When AI is used, it makes possible for machines to use the experience for learning. Once it collects enough experiences, it is capable of producing the output for the new set of inputs in the similar way a human does. AI is a wide scientific field as the intelligence can be applied in various ways. Two related scientific fields which are closely related to AI are statistics and computer science. This is obvious, since one needs statistics in order to define and describe associated algorithms, and computer science is needed to translate the algorithms into a machine language in order to perform actions. Figure 4.1 shows the position of AI in relation to computer science and statistics with related applications.

ML is one of the main components of AI. It is a set of learning computer algorithms by which machines or computers can learn without explicitly being programmed. Moreover, ML can be defined as the field of AI which provides the algorithms for machines to automatically learn and improve their actions from a

B. Hrnjica (✉)
University of Bihac, Bihac, Bosnia and Herzegovina
e-mail: bahrudin.hrnjica@unbi.ba

A. Danandeh Mehr
Antalya Bilim University, Antalya, Turkey
e-mail: ali.danandeh@antalya.edu.tr

© Springer Nature Switzerland AG 2020   71
F. Al-Turjman (ed.), *Smart Cities Performability, Cognition, & Security*,
EAI/Springer Innovations in Communication and Computing,
https://doi.org/10.1007/978-3-030-14718-1_4

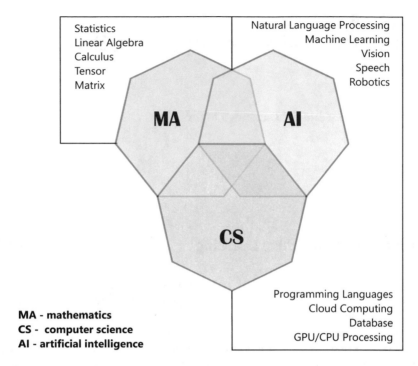

Fig. 4.1 AI as scientific filed in relation with statistics (mathematics) and computer science

given experience. While AI is defined as the ability to acquire and apply knowledge, ML is defined as the acquisition of knowledge or skill. On the other hand, using AI one tries to increase the chance of success but not accuracy. However, in ML one tries to increase the accuracy of the action regardless of the success. Last but not least, AI can be defined as a smart computer program, while ML is the concept of how machines use data to learn. As stated previously, ML is a set of computer algorithms, particularly designed for machines to help them in the learning process. Usually, an ML process consists of searching the data to recognize hidden patterns in the data. Once the patterns are recognized, the computer can make predictions for new or unseen data based on persisted knowledge. Supervised, unsupervised, and reinforcement learning are basically the three main types of the ML (Fig. 4.2):

In supervised ML, the learning process consists of finding the rule that maps inputs (features) to outputs (labels). During the learning process, available data can be divided into two sets. The first set is the training set which is responsible for the training process. The second set is called validation or testing set and is used by the learning algorithm to verify the training process.

Unsupervised learning is the process of discovering patterns in data without defined output (unlabeled dataset). With unsupervised learning, the correct result cannot be determined because no output variable is defined. Algorithms are left to their capability to discover as much knowledge as possible from the data. Inasmuch

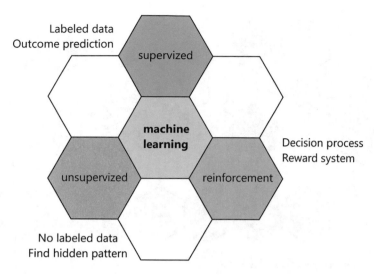

Labeled data
Outcome prediction

supervized

machine
learning

Decision process
Reward system

unsupervized

reinforcement

No labeled data
Find hidden pattern

Fig. 4.2 Machine learning types

as there is no output variable, there is no need for splitting the data into training and testing sets. Thus, all the samples are used for training process. This kind of learning can be applied in image and signal processing, computer vision, etc.

Reinforcement learning is an ML process where a computer interacts with a dynamic system in which it must achieve a goal (like driving a vehicle or playing a game). Reinforcement learning provides feedback that consists of information showing how the last action was treated, was it successful or unsuccessful. Based on the feedback, the computer can learn and make further decisions.

Supervised ML can be classified by the type of the output as regression or classification. In regression the output is represented as a continuous number, while in classification, the output variable is discrete rather than continuous, and consists of two or more classes.

Regression ML considers finding relations between one or more variables which are called features, and then compared with a dependent variable called the label. Figure 4.3 shows how ML can be classified depending on its learning types.

ML-originated regression models are used in different disciplines such as finance, production, the stock market, and maintenance. The models can later be used for predicting or forecasting sales, weather temperature for the next day, stock market prices in the next few hours, energy consumption for a given time period, etc.

In most cases, a dataset is determined by time so that each dataset value has a defined timestamp. In this way the history of a dataset value can be monitored, which can be valuable for future business decisions. This kind of dataset is called time series data. In one example, a stock price is a set of observed values recorded in time, so for each time value (minutes, hours, or days) a stock price is calculated. In a second example daily energy consumption is collected so that at the end of each

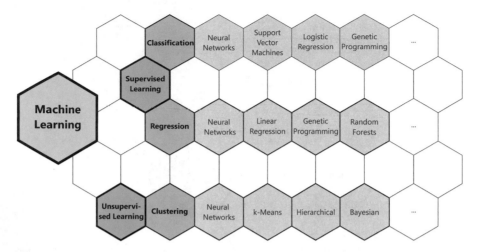

Fig. 4.3 Different type of ML algorithms

day it can be recorded. Regression type of ML covers broad applications. One of the most complex types of ML are time series events.

The forecasting of time series events is an important area of ML, since there are so many forecasting tasks that involve a time component. The time component adds additional information to the time series data, but it also makes time series problems more difficult to handle in comparison to many other regression tasks. Predictions of the time series events are still one of the most difficult tasks, and are active research subjects of many engineers and scientists. Time series events can be detected around us, and the prediction of its future states is of tremendous importance. For example, it is of crucial importance for the world economy to forecast the price of energy [25], sales [13], or stock prices [17]. Furthermore, time series events can be used to predict the weather and environmental, hydrological, and geological events [4, 6, 18]. Smart cities need better and smarter surveillance cameras [22], digital surveillance systems with different frameworks [23, 32], and 5G-inspired IIoT paradigm in health care [1].

In this chapter, the application of deep learning is used in order to present an approach of forecasting energy demand that can be part of a smart cities cloud solution. Since our cities face non-stop growth in population and infrastructures, handling its resources in an intelligent way may result in multiple cost savings. One of the resources which is very important for smart cities is electricity, it is of crucial importance that it is handled in an efficient and intelligent way. The basic role of the smart energy concept is to optimize its consumption and demand resulting in decreased energy costs and increased efficiency. Among the variety of benefits, the smart energy concept mainly enhances the quality of life of the inhabitants of the cities as well as making the environment cleaner. One of the approaches for the smart energy concept is to develop prediction models using ML algorithms in order to forecast energy demand, especially for daily and weekly periods.

The upcoming chapter describes thoroughly what is behind the deep learning concept as a subset of ML and how neural networks can be applied for developing energy prediction models. A specialized version of the RNN, e.g., LSTM, and unsupervised neural network type called autoencoders are described in detail. With an autoencoder unsupervised neural network, features are transformed so that important information is not lost due to high inter-correlation between them. With LSTM the historical influence of the data has been captured. The LSTM can capture the long-time dependency with constant error propagation while using backprop-agation through time BPTT, which outperforms the standard implementation of the RNN. To build a deep learning model, the computer program ANNdotNET is introduced. The ANNdotNET is an open source project hosted at https://github.com/bhrnjica/anndotnet on GitHub, the largest open source repository platform. The ANNdotNET provides a user-friendly ML framework with the capability of importing data from the smart grids of a smart city. By design, the ANNdotNET is a cloud solution program that can be connected with other IoT devices for data collecting, feeding, and providing efficient models to energy managers for a bigger smart city cloud solution. As an example, the chapter provides the evolution of daily and weekly energy demand models for Nicosia, the capital of Northern Cyprus. Currently, energy demand predictions for the city are not as efficient as expected. Therefore, the results of this chapter can be used as efficient alternatives for IoT-based energy prediction models for the city.

Using prediction models based on deep neural network, one might not be able to answer questions about behavior, seasonality, and trends of a given time series. To cope with this problem, two state-of-the-art time series decomposition algorithms are used here in order to analyze and determine the trend and seasonality of energy demand. Moreover, the prediction model based on time series decomposition called TSD has been developed and compared with the deep learning model.

Artificial Neural Network

A number of ML algorithms have been developed over the past few decades that try to discover knowledge from the data. Which ML algorithm is the best, for a given problem? Is the algorithm satisfactory? These are the two main questions that were discussed in the thousands of ML studies. It seems that the former attracted less attention than the latter. Artificial neural network, ANN, undoubtedly is one of the most popular ML algorithms that every data scientist has heard about it. It is a part of supervised ML algorithms that is based on the concept of the biological neural network. Similarly, as a genetic algorithm GA tries to mimic biological evolution [15], ANN attempts to simulate the decision process as human neurons do. The concept of ANN is based on the neuron that can be described as the basic cell of the human brain. Each neuron consists of a cell, a tubular axon, and dendrites. The cell processes signals coming from dendrites and sends it to the axon. The axon forms synaptic connections with other neighboring neurons. The axon consists of branched ends which are used as the input for the next neuron cell. Neurons are linked via a synapse where signals are exchanged from one neuron to another.

Akin to the biological neuron, the artificial neuron is defined as a set of input parameters x_i ($i = 1, .. n$), which represents the input signals, set of weight factors w_i ($i = 1, ..., n$), which represents the synapses, the dot product $\sum \mathbf{w} \cdot \mathbf{x}$ of input and weighted vectors, representing the neuron cell, and activation function a (.), representing the axon of biological neurons [20]. Figure 4.4 shows the similarities between the biological and artificial neuron.

The first concept of the artificial neuron is called perceptron which was introduced by Rosenblatt [26]. Let the x_n represent the input vector with n components, the associated weight w_n, and bias value b_0 and activation function $sign$. The output y of the perceptron can be expressed as:

$$y = f\,(net) = f\,(\mathbf{w} \cdot \mathbf{x}) = sign \left(\sum_{i=1}^{n} x_i w_i + b_0 \right), \qquad (4.1)$$

where sign represents the activation function defined as:

$$sign\,(net) = \begin{cases} +1, if\ net \geq 0, \ \mathbf{w} \cdot \mathbf{x} \geq 0 \\ -1, if\ net < 0, \ \mathbf{w} \cdot \mathbf{x} < 0 \end{cases} \qquad (4.2)$$

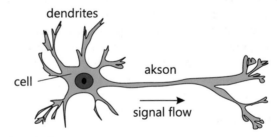

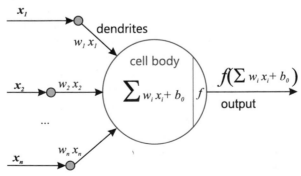

Fig. 4.4 Graphic interpretations of biological and artificial neurons and their similarities

As can be seen the activation function is the last operation in the expression (4.2), which obviously shows the perceptron produces an output as binary value of 1 and −1.

Besides *sign* there are many other activation functions which can produce different kind of outputs, e.g., *Tanh*, *ReLU*, *Sofmax*, etc. The expression (4.1) and (4.2) with any combination of the activation functions represents the forward pass of one neuron. The forward pass calculates the output for a given input and weights values. In context of a whole ANN, forward pass calculates the output of each neuron in the network, where the last neuron's output represents the output of the network.

Just as millions of biological neurons can be connected, the artificial neurons can form an ANN which can solve very complex problems. Neurons in ANN are grouped in layers. Each layer can consist of one or more neurons. Usually, ANN layers are classified as input, output, and hidden layers. The input layer represents the layer constructed from the input variables (called features). The number of neurons in the input layer is always related by the number of features. Similarly, the output layer is based on the output variable. The number of output variables (called labels) must be the same as the number of neurons in output layer.

The simplest ANN can be formed from at least one input, one hidden, and one output layer. This simple network configuration is called the feed forward network, FFN. The number of neurons of each layer may vary depending on the complexity of the problem. In the input layer, each neuron corresponds to the input parameters, while the output layer is related to the output result. In the middle of the input and the output layer there can be one or more hidden layers with arbitrary numbers of neurons. Figure 4.5 shows the FFN with four layers, one input, one output, and two hidden layers.

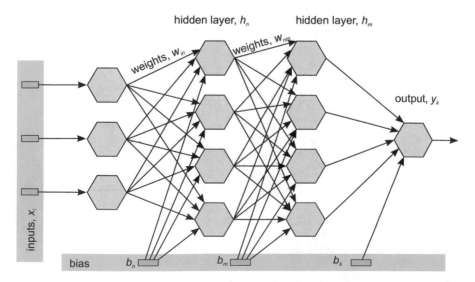

Fig. 4.5 FNN with four interconnected layers

Since each neuron makes some kind of decisions, one can conclude that one perceptron cannot do much. In case of the ANN previously described, the ANN can produce valuable decisions, which can lead to a solution to complex problems. In all cases the process of producing a better solution depends on the weighted factor values of each neuron.

Learning Process in ANN

The process of finding suitable weight values is called learning of ANN. Finding the weight values starts with output calculations of each neuron in the network. The process begins with output calculations of neurons in the input layer, then the output of each input neuron becomes the input for the neurons in the first hidden layer, and so forth. Once the result of the last neuron is calculated in the output layer, the result becomes the output of the network. The result is calculated for each sample (row) in the training dataset.

Assume the training dataset is defined with two features and one label. Let Table 4.1 represent the training dataset with 3 samples (rows). As illustrated in Fig. 4.6, FFN with a 2-4-1 structure was used and indicates that the input layer

Table 4.1 Sample training dataset consists of two features (X_1 and X_2) and one label (Y), with three data samples (rows)

X_1	X_2	Y
1	1	2
1	2	3
3	1	4

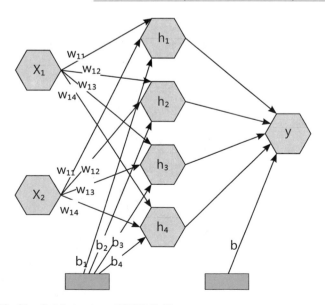

Fig. 4.6 FNN with a 2-4-1 structure, FNN(2, 3, 1)

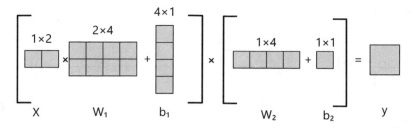

Fig. 4.7 Matrix multiplication generated from the previously defined FNN (2, 3, 1)

has two neurons, the hidden layer has four neurons, and the output layer has one neuron. Identity activation function ($x = f(x)$) at both hidden and output layers was applied for the sake of simplicity.

Once the training dataset and network configuration are defined, the network output can be calculated. The ANN output calculation is based on the matrix calculation. Figure 4.7 shows a matrix representation of the network given in Fig. 4.6.

Based on Fig. 4.7, and the training dataset given in Table 4.1, the output is calculated for each row of the datasets, but first, the initial values of the weights and biases must be defined. This is usually a random process. Assume that the following values are assigned to the matrices W_1 and W_2, and biases b_1 and b_2.

$$W_1 = \begin{bmatrix} 0.5 \ 0.5 \ 0.5 \ 0.5 \\ 0.6 \ 0.6 \ 0.6 \ 0.6 \end{bmatrix}; \quad W_2 = \begin{bmatrix} 1.8 \ 0.4 \ 0.4 \ 0.4 \end{bmatrix} \tag{4.3}$$

$$b_1 = \begin{bmatrix} 0.1 \\ 0.12 \\ 0.10 \\ 0.11 \end{bmatrix}; \quad b_2 = \begin{bmatrix} 0.1 \end{bmatrix} \tag{4.4}$$

Now, the network output can be calculated as:

$$[X \cdot W_1 + b_1] \cdot [W_2 + b_2] = [\hat{y}], \tag{4.5}$$

Using the data from Table 4.1, the outputs are calculated for each row:

- Row 1:

$$\hat{y}_1 = \left[\begin{bmatrix} 1 \ 1 \end{bmatrix} \begin{bmatrix} 0.5 \ 0.5 \ 0.5 \ 0.5 \\ 0.6 \ 0.6 \ 0.6 \ 0.6 \end{bmatrix} + \begin{bmatrix} 0.5 \\ 0.5 \\ 0.5 \\ 0.5 \end{bmatrix} \right] \cdot \begin{bmatrix} \begin{bmatrix} 1.8 \ 0.4 \ 0.4 \ 0.4 \end{bmatrix} + \begin{bmatrix} 0.4 \end{bmatrix} \end{bmatrix} = 2.26.$$

$$\tag{4.6}$$

- Row 2:

$$\hat{y}_2 = \left[[1\ 2] \begin{bmatrix} 0.5\ 0.5\ 0.5\ 0.5 \\ 0.6\ 0.6\ 0.6\ 0.6 \end{bmatrix} + \begin{bmatrix} 0.5 \\ 0.5 \\ 0.5 \\ 0.5 \end{bmatrix} \right] \cdot [[1.8\ 0.4\ 0.4\ 0.4] + [0.4]] = 3.34.$$

(4.7)

- Row 3:

$$\hat{y}_3 = \left[[3\ 1] \begin{bmatrix} 0.5\ 0.5\ 0.5\ 0.5 \\ 0.6\ 0.6\ 0.6\ 0.6 \end{bmatrix} + \begin{bmatrix} 0.5 \\ 0.5 \\ 0.5 \\ 0.5 \end{bmatrix} \right] \cdot [[1.8\ 0.4\ 0.4\ 0.4] + [0.4]] = 4.06.$$

(4.8)

Based on the expressions (4.6), (4.7), and (4.8) the predicted value represents the column vector of three elements:

$$\hat{y} = \begin{bmatrix} 2.26 \\ 3.34 \\ 4.06 \end{bmatrix}.$$

(4.9)

On the other hand, the actual outputs from Table 4.1 can be represented by the column vector as:

$$y = \begin{bmatrix} 2 \\ 3 \\ 4 \end{bmatrix}.$$

(4.10)

The residual vector e, which is the deference between the actual and predicted values, is given as:

$$e = \begin{bmatrix} 2 - 2.26 \\ 3 - 3.34 \\ 4 - 4.06 \end{bmatrix} = \begin{bmatrix} -0.26 \\ -0.34 \\ -0.06 \end{bmatrix}.$$

(4.11)

The learning process is based on defining the cost function C which can be of various types depending on the problem at hand. In most cases, it is the squared error between the actual and predicted results [27]:

$$C = \frac{1}{2} \sum_{i=1}^{n} e_i^2,$$

(4.12)

where:

- e is the residual values,
- n is the number of data samples for the training,
- y is the actual values,
- $\hat{y}$ - is the calculated values.

In case of the previous example, the cost function produces the following error value:

$$C = \frac{1}{2} \sum_{i=1}^{3} (y_i - \hat{y_i})^2 = \frac{1}{2} \left((-0.26)^2 + (-0.34)^2 + (-0.06)^2 \right) = 0.0934.$$

(4.13)

Minimizing the error value is the central part of training process with an ANN and might result in satisfactory predictions. One common way is to apply the backpropagation algorithm that is the iterative process of correcting weights and biases based on the partial derivative of the cost function.

For instance, at each neuron k, the weight value $w_k{}^i$ at the iteration i will have the new value $w_k{}^{i+1}$ in the next iteration $i + 1$, for the gradient of the cost function with respect to the w_k multiplied with the learning rate factor η:

$$\Delta w_k = \eta \, \frac{\partial C}{\partial w_k}.$$

(4.14)

The new value of the weight, w_k, in the $i{+}1$ iteration is expressed as:

$$w_k{}^{i+1} = w_k{}^i + \Delta w_k.$$

(4.15)

For the calculation of the cost function, the gradient starts from the last (output) layer and is propagated backwards to the input layer using the chain derivative rule.

The entire learning process can be described in two stages for each iteration. Once the iteration starts, the output is calculated by starting from the input layer, and for each weight and each input variable, the output is calculated for each neuron. Once the output of the network is calculated, the second process is started by calculating the gradient from the output layer to the input layer in the backwards order. For each weight value, the gradient is calculated and added to the previous values as shown in Eq. (4.15). Training of the FFN can be a complex task, since not all ANN models can solve the problem accurately. Since an ANN model can be built with one or more hidden layers, and each hidden layer can contain an arbitrary number of neurons, the learning process may provide unexpected results. In the case of small number of neurons in the hidden layer, the model may be too rigid, and the learning process very slow, which leads to the fact that the number of neurons in the ANN is not sufficient to adapt to the data. On the other hand, a large number of neurons in the hidden layer may lead the ANN model to fit the data perfectly, but due to the complex nature of the problem the model is trying to predict, the prediction for the unseen data may give unsatisfactory results [3].

Deep ANN

Previous studies indicate that FFNs are not powerful enough for most of today's problems. For instance, FFNs were not found suitable for the natural language processing, social network filtering, speech and audio recognition, machine translation, medical image analysis, bioinformatics, drug design, stochastic time series forecasting, etc. In order to solve such problems, the network configuration must be extended and made more robust. To achieve a more robust network, one may increase the number of hidden layers. In this situation, the multiple hidden layers with a nonlinear activation function can produce nonlinear processing which is more efficient for solving complex problems. Simply by increasing the number of hidden layers produces a very complex system of network configurations that need to be learned. In most cases more than one hidden layer in the network cannot be learned in the same way described previously, due to the vanishing and exploding gradients phenomenon [2]. This changes the approach of looking at ANNs and revolutionized the learning process for multiple layer networks.

As stated previously the keyword "deep" in "deep ANN" refers to the number of hidden layers. One can define the credit assignment path (CAP) depth as the transformation chain from the input to output of the neural network [30]. In the case of FNN, the depth of the CAP indicates the number of hidden layers plus one for the output layer, since it is parametrized in the same way as the hidden layer. For RNN where the input features can be propagated through a layer more than once per iteration, the CAP depth is potentially undetermined. So, to make a measurable difference between deep learning and learning requires a CAP depth to be greater than 2. The process of learning complex networks configuration, such as deep neural networks, deep belief networks, and RNN, is called deep learning, DL, a subset of the wider ML field. How DL is specific to the ML field can be depicted graphically as in Fig. 4.8. The learning process of a DL specific networks is a very complex task, it is still based on the backpropagation error concept with a specific way of error propagation and optimization techniques.

Recurrent Neural Network

FFN models are usually built on the fact that data do not have any order when entering into the network. So, the output of ANN depends only on the input features. In case of specific data when the order is important, usually when data is recorded in time or when dealing with sequences of data, simple FFN cannot manage it as one can expect because the previous state cannot be incorporated [24]. When the output

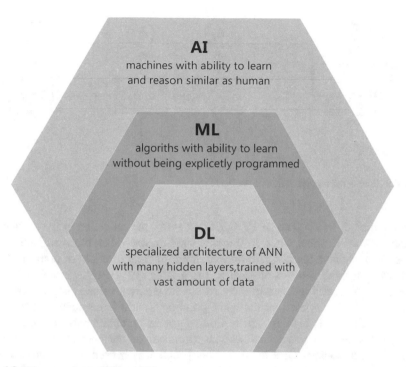

Fig. 4.8 DL as a subset of ML and AI

is determined by both the inputs and the previous states, the FNN must be extended to support the previous states. A well-known solution for this kind of problem is to develop the RNN, which was first introduced by Hopfield [12], and later popularized when the backpropagation algorithm was improved [24]. The concept of the RNN is depicted in Fig. 4.9. As seen, the RNN contains cycles showing that the current state of the network relies on current data, but also on the data produced by the previous outputs of the network. So, in the case of the RNN, two kinds of inputs are provided: the output of the previous time, h_{i-1}, and the current input x_i. Due to its nature, the RNN has a special kind of internal memory which can hold long-term information history [29].

Figure 4.9 shows two kinds of representations of the RNN. On the left side, the RNN is presented in classic feed forward like mode, where the three layers are presented: input, hidden, and output layer. Around the hidden layer we can see cycling which indicates the recursion. The RNN can be shown in an unrolled state in time t. The RNN is presented with t interconnected FFN, where t indicates the past steps, thus far. The concept of the RNN is promising and very challenging, but there are problems with applications, mainly when dealing with complex time dependent models [2]. Most of the obstacles of the RNN can be summarized in two categories [2]: the vanishing and exploding gradient. The learning process of the RNN is mostly based on the backpropagation algorithm, the so-called backpropagation

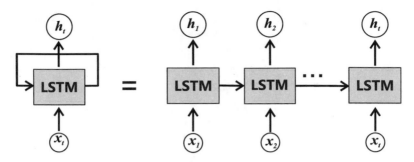

Fig. 4.9 Schematic representation of the RNN

through time or BPTT. The BPTT algorithm stores the activation of the units while going forward in time, while in the backwards phase takes those activations for the gradient calculation [27]. In the vanishing gradient problem of learning RNN, updates of weights are proportional to the gradient of the error calculated in the previously described manner. In most cases, the gradient value is negligibly small, which results in the fact that the corresponding weight is constant and stops the network from further training. The exploding gradient problem refers to the opposite behavior, where the updates of weights (gradient of the cost function) become larger in each backpropagation step. This problem is caused by the explosion of the long-term components in the RNN. In both cases the error propagation through the network is not constant, which causes one of the mentioned problems.

The solution to the abovementioned problems is found in the specific design of the RNN, called long short-term memory, LSTM [11]. LSTM is a special RNN which can provide a constant error flow. The constant error propagation through the network involves a special network design. LSTM consists of memory blocks with self-connection defined in the hidden layer, which has the ability to store the temporal state of the network. Besides memorization, an LSTM cell has special multiplicative units called gates, which control the information flow. Each memory block consists of the input gate that controls the flow of the input activations into the memory cell, and the output gate controls the output flow of the cell activation. In addition, an LSTM cell also contains the forget gate, which filters the information from the input and previous output and decides which one should be remembered or forgotten and dropped. With such selective information filtering, the forget gate scales an LSTM cell's internal state, which is self-recurrently connected by previous cell states [8]. Besides gating units, the LSTM cell consists of a self-connected linear unit called constant error carousel, CEC, whose activation is called the cell state. The cell state allows for constant error flow, previously mentioned as the problem of the vanishing or exploding gradient, of the backpropagation error in time. The gates of the LSTM are adaptive, since each time the content of the cell is out of date, the forget gate learns to reset the cell state, so the input and the output gates control the input and the output, respectively. Figure 4.10 shows an LSTM cell with activation layers: input, output, forget gates, and the cell. Each layer contains the activation function before passing through.

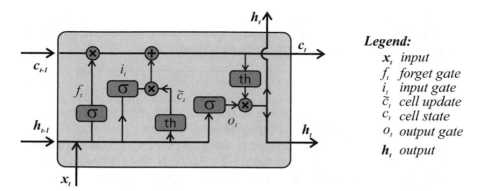

Fig. 4.10 LSTM cell with its internal structure

As can be seen, an LSTM network can be expressed as an ANN where the input vector $x = (x_1, x_2, x_3, \ldots x_t)$ in time t maps to the output vector $y = (y_1, y_2, \ldots, y_m)$, through the calculation of the following layers:

- the forget gate sigmoid layer for the time t, f_t is calculated by the previous output h_{t-1}, the input vector x_t, and the matrix of weights from the forget layer W_f with an addition of bias b_f:

$$f_t = \sigma \left(W_f \cdot [h_{t-1}, x_t] + b_f \right); \qquad (4.16)$$

- the input gate sigmoid layer for the time t, i_t is calculated by the previous output h_{t-1}, the input vector x_t, and the matrix of weights from the input layer W_i with an addition of bias b_i:

$$i_t = \sigma \left(W_i \cdot [h_{t-1}, x_t] + b_i \right); \qquad (4.17)$$

- the cell state in time t, C_t, is calculated from the forget gate f_t and the previous cell state C_{t-1} by multiplicative operation $\otimes$. The result is applied as the first argument of the additive operation $\oplus$ and the input gate i_t, which is then applied as the first argument of the multiplicative operation of the cell update state $\tilde{c}_t$ which is a *tanh* layer calculated by the previous output h_{t-1}, input vector x_t, and the weight matrix for the cell with an addition of bias b_C:

$$C_t = f_t \otimes C_{t-1} \oplus i_t \otimes \tanh (W_C \cdot [h_{t-1}, x_t] + b_C); \qquad (4.18)$$

- the output gate sigmoid layer for the time t, o_t is calculated by the previous output h_{t-1}, the input vector x_t, and the matrix of weights from the output layer W_f with an addition of bias b_o:

$$o_t = \sigma (W_0 \cdot [h_{t-1}, x_t] + b_0) . \qquad (4.19)$$

The final stage of the LSTM cell is the output calculation of the current time h_t. The current output h_t is calculated with the multiplicative operation $\otimes$ between output gate layer and $tanh$ layer of the current cell state C_t.

$$h_t = o_t \otimes \tanh(C_t). \tag{4.20}$$

The current output, h_t, has passed through the network as the previous state for the next LSTM cell, or as the input for an ANN output layer. The operation connections $\otimes$ and $\oplus$, which correspond to multiplication and addition connections, allow the gates to process the information based on the previous cell output, as well as the previous cell state. The previous LSTM cell description represents one of the several variants which can be found in the literature [19].

Deep LSTM RNN

Deep ANN has proved to be very effective in solving complex problems. Similarly, deep LSTM RNN can be defined as more than one LSTM layer in the ANN. The fact is that an unrolled LSTM cell in time represents a deep FFN, which indicates the complex network architecture. Deep LSTM RNN can be defined with more than one LSTM layer. The LSTM layers are stacked vertically, with the output sequence of one layer forming the input sequence of the next one (Fig. 4.11).

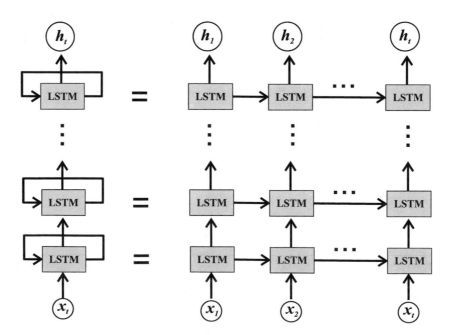

Fig. 4.11 Schematic representation of the deep LSTM RNN

The deep LSTM RNN has proved to be very effective and outperforms the standard implementation of the LSTM [9]. Deep LSTM RNN has the ability to learn at different time scales over the input [10]. In addition, deep LSTM is more efficient than standard LSTM, since parameter distribution over the space through multiple layers is handled with less memory [29].

Modeling Time Series Events

As previously mentioned, ML methods generally extract the knowledge from the data in two phases. The first phase is training the network configuration in order to get the best possible weights so that the model can predict the future values with minimum error. The second phase is model validation, where the trained model is validated against the validation dataset. Such dataset contains the data which are not used during the training phase. In both of the phases, the data is a crucial part of the ML solution. With noisy or inappropriate datasets, regardless of the implemented ML algorithm, a reliable prediction model would not be created. Therefore, the most important component of the ML solution is to prepare high quality datasets prior to the training phase.

Undoubtedly, modeling time series data is one of the most challenging tasks of ML. A time series represents a sequential set of data samples recorded over successive times. It can also be defined as a set of vectors $x(t)$, $t = 1, 2, 3, \ldots$, where t is the time. The time series data is always arranged in chronological order. Data usually contains a single variable, which represents the univariate time series. In case more than one variable is used, it is termed as a multivariate time series.

For better understanding time series events, the associated data can be decomposed into several components so that each component represents an important property of the event. Decomposition of the time series is usually based on rates of change. With this in mind, time series can be decomposed into three components: trend, seasonal, and random components [16]. As an underlying component, trend represents tendency of time series data to increase, decrease, or stagnate over a long period of time. It can be also described as long-term movement. The seasonal component represents the fluctuation within a year. The seasonal variation of time series is an important component specially in business related time series data, where the detection of a seasonal time interval can increase business values. The third component, the random component, represents everything else. It usually represents the randomness of the time series.

The time series components can be determined in two ways: as additive or as multiplicative models. In case of the additive model, the time series data can be expressed as:

$$y(t) = S(t) + T(t) + R(t) \tag{4.21}$$

where $y(t)$ is the data, $T(t)$ is the trend component, $S(t)$ is the seasonal component, and the $R(t)$ is the random component at time t.

The multiplicative model of time series can be expressed as:

$$y(t) = S(t) \times T(t) \times R(t) \tag{4.22}$$

In case when the seasonal variation is relatively constant over time, additive decomposition is recommended. On the other hand, the multiplicative decomposition is recommended when the seasonal component is proportional to the level of the time series [16], or when the seasonal variation increases over time. However, the multiplicative model can be expressed as additive, if the time series is transformed by a log transformation. In this case, using a log transformation any multiplicative decomposition can be expressed as additive:

$$y(t) = S(t) \times T(t) \times R(t) \Leftrightarrow Ln(y(t))$$
$$= Ln(S(t)) + Ln(T(t)) + Ln(R(t)). \tag{4.23}$$

In order to decompose time series data into its components several methods have been developed. One of the most popular methods is seasonal-trend decomposition based on Loess [5], which has been proven to be very effective for longer time series. There is a more recent seasonal-trend decomposition based on regression decomposition [7], which is more generic, and it allows for multiple seasonal and cyclic components, as well as multiple linear regressors with a constant. It also provides flexible, seasonal, and cyclic influences. Recently, Facebook has released a times series decomposition and forecasting tool called Prophet. The company used this tool for times series data analytics and forecasting [31]. All the mentioned decomposition methods are implemented in well-known statistical packages, e.g., R or Python languages.

Time Series Decomposition Procedure

In order to decompose the time series, the first step is the selection of a decomposition method. Once the decomposition method has been selected, the trend is the first component which should be estimated. De-trending the time series is the second step. In case of an additive model, the trend component is subtracted from the time series data. For the multiplicative model, time series data is divided by the estimated trend component.

The seasonal component is estimated from the de-trended part of the time series. Since the seasonal component is based on the underlying periodicity of the events, it can be weekly, monthly, yearly, or any customized seasonal length. One of the simplest methods to estimate, the seasonal component is to average the de-trended values for the specific season. For example, to get a seasonal effect for January, one can average the de-trended values for each January in the series. The seasonal value has to be adjusted depending of the decomposition type, zero for additive, and one for a multiplicative model.

The final step of the decomposition is to estimate the random component. The random component is simply estimated when the trend and season are subtracted from the data series. In case of the multiplicative decomposition, the random component is estimated by dividing the multiplication of trend and seasonal components by the time series data.

The following expressions for the random component can be stated as:

$$R(t) = y(t) - [S(t) - T(t)] - \text{for additive decomposition}, \tag{4.24}$$

$$R(t) = \frac{y(t)}{S(t) \times T(t)} - \text{for multiplicative decomposition}, \tag{4.25}$$

Time series decomposition with its components is very often graphed, since it provides a clear picture for the data behavior.

Energy Demand Analysis by Time Series Decomposition

Reliable electricity prediction models lead to a sustainable power supply and provide clear details about the health of a power system in a smart city setup. In this section, we demonstrate how a decomposed time series of energy consumption data can be used to develop accurate electricity demand prediction models. As an example, the daily series of electricity consumption in the northern part of Nicosia during the 2011–2016 period has been considered. Nicosia, the capital city of Cyprus, has a typical Mediterranean climate with an annual average electricity consumption about 4000 MWh in its northern part and about 6000 MWh in the southern part. The overall image of the electricity consumption time series is shown in Fig. 4.12. Based upon observations, the rapidly increasing trend has been observed since 2013. The figure also represents seasonal behavior, since the sinusoidal shape of the data can be visually recognized. The statistical properties of the electricity consumption series are shown in Table 4.2.

Quantile values indicate that the data is skewed to the left since the median has a lower value than the mean.

The decomposition of the dataset was performed by using the STL R package [5]. In order to get the best possible decomposition with maximum amplitude values of the seasonal component, the decomposition was performed for wide range of argument values. The best possible decomposition was estimated for the argument value of $f = 365$, which indicates yearly periodic behaviors of the dataset. Figure 4.13 shows the decomposition of the energy demand time series.

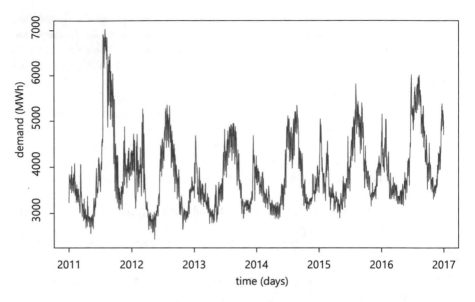

Fig. 4.12 Energy demand time series at northern part of Nicosia

Table 4.2 Summary of electricity consumption 2011–2016

Min	1st quarter	Median	Mean	3rd quarter	Max
2433	3311	3664	3879	4284	7032

Figure 4.13 also indicates that the random component has significant influence on the energy demand dataset. From Table 4.3, one can see that the basic statistical indicators show that the minimum value is less than −900 MWh, while the maximum value is greater than 1700 MWh. This implies that the forecasting model may produce 50% of the relative error.

The seasonal component can be described as a periodic function with two minimum and one maximum points. The first minimum point is reached in 122 days, the second week in May, while the second minimum point is reached in the first week of November. The maximum energy demand is reached in the first week of August. The maximum point of seasonal change is expected in the beginning of August because of high electricity demand due to high temperatures and increased tourisms. The trend component of the energy demand dataset shows growth in the last 4 years. This indicates that energy demand increases every year. The reason for the increasing trend can be found in the constant growth of infrastructure and population on the island.

From the previous decomposition time series, over the last years, one can conclude that energy demand has constant growth. Over a period of a year, the

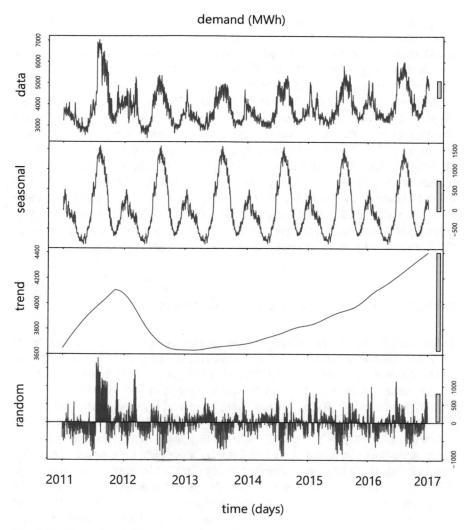

Fig. 4.13 Energy demand time series components

Table 4.3 Summary of random component of energy demand time series in MWh

Min	1st quarter	Median	Mean	3rd quarter	Max
−928.684	−201.329	−21.750	2.466	152.537	1776.062

energy demand reaches two minimum levels: one in the spring and one in the fall. The maximum energy demand is reached during the first days of August, due to maximum temperatures. Seasonal and random components have similar ranges that indicate the energy demand has strong stochastic behavior.

In order to get deeper into seasonal changes, one should see how demand changes weekly, monthly, and quarterly. In order to display different seasons, the Prophet *R* package is used [31]. Figure 4.14 shows the trend and four different seasonal components: weekly, monthly, quarterly, and yearly. The yearly component in Fig. 4.14 clearly shows two minimum and one maximum point, which were previously described. Weekly seasonal changes indicate that the energy demand is higher on work days rather than on the weekend. The monthly seasonal changes do not clearly show increasing or decreasing demand, but it roughly indicates that energy demand is on the highest level in the middle of the month. Similarly, quarterly seasonal changes are increased at the start and the end of the quarter, while the first month of the quarter shows gradual decreased demand.

Energy Demand Forecasting Using Decomposed Series

In the previous section, the energy demand time series was analyzed by decomposing it into three main components: trend, seasonal, and random. Moreover, different types of seasonality were considered in order to get a deeper knowledge of the energy demand data. Based on the previous analysis in this section, forecasting will be performed in order to see how energy demand is propagated in the future. The forecasting procedure was performed using the additive model where nonlinear trends are fit with yearly, weekly, and monthly seasonality. Since the seasonal components have significant impact on the time series data, the prophet forecasting package is used. The package combines many different forecasting methods (e.g., ARIMA, exponential smoothing, etc.) in order to get the best possible model. The forecasting is based on using a flexible regression model or curve-fitting model instead of a traditional time series model that leads to better and more accurate forecasting. The time series decomposition, TSD, model is built on the energy demand dataset from January 2011 to September 2015. From October 2015 to December 2016 the dataset is defined as the validation and testing dataset.

Figure 4.15 shows the TSD model of the energy demand time series data. The black dots are actual values of the daily energy demand, whereas the blue (dark gray) line shows predicted values. Besides prediction values, the image also shows the confidence interval of the prediction.

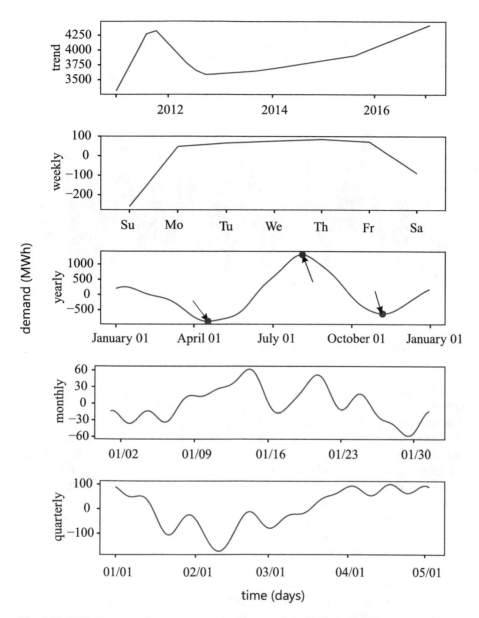

Fig. 4.14 Different seasonal component types of energy demand time series data

LSTM Deep Learning Model for Energy Demand Prediction

Using a deep learning technique for modeling time series events seems natural due to the complex nature of such events. In this section, a deep learning model has been

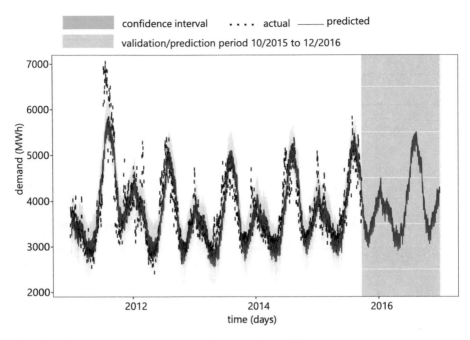

Fig. 4.15 TDS model of energy demand for period 2011–2016

developed in order to predict energy demand. Since the energy demand represents a typical time series non-stationary dataset, the LSTM RNN was used. In order to transform the time series into a data frame, 15 days of time lags were used. Once the data frame was created, the three datasets were created to configure the ML workflow. In order to transform the data, configure the neural network, train and evaluate the model, the ANNdotNET [14]—deep learning tool on .NET platform was used. The ANNdotNET is a deep learning tool that implements the ML Engine that is based on the Microsoft Cognitive Toolkit, CNTK [34]. The ML Engine is responsible for training and evaluating deep learning models. Besides the GUI tool that is used for handling data transformation and model training, the ANNdotNET provides a set of APIs which could be integrated into a bigger smart cities cloud solution, SCCL.

The previous section is correlated for data analysis, where the time series data were decomposed and analyzed. The time series is represented with only one variable (energy demand) which is the example of a univariate time series. In order to prepare it for deep learning, it must be transformed into a data frame-based set, with features (input) and label (output). The features are generated by the previous values, so-called time lag values, while the label is the time series value at the current time step. Figure 4.16 shows how a univariate time series can be transformed into a data frame with 15 features and one label. In this way, the time series data is transformed so that historical changes have an influence on the current value. For

Date	Mwh
01-Jan-11	3235
02-Jan-11	3474
03-Jan-11	3840
04-Jan-11	3784
05-Jan-11	3708
06-Jan-11	3680
07-Jan-11	3702
08-Jan-11	3714
09-Jan-11	3525
10-Jan-11	3738
11-Jan-11	3740
12-Jan-11	3777
13-Jan-11	3845
14-Jan-11	3777
15-Jan-11	3570
16-Jan-11	**3448**
17-Jan-11	4021
18-Jan-11	3760
19-Jan-11	3646
...	...
31-Dec-16	4706

Date	MWh-15	MWh-14	...	MWh-2	MWh-1	Mwh
16-Jan-11	3235	3474	...	3777	3570	**3448**
17-Jan-11	3474	3840	...	3708	3680	4021
18-Jan-11	3840	3784	...	3680	3702	3760
19-Jan-11	3784	3708	...	3702	3714	3646
20-Jan-11	3708	...	...	...	...	...
...	...	...	...	...	...	...
...	...	...	...	...	...	...
31-Dec-16	4994	5060	...	4839	5024	4706

Fig. 4.16 Time series transformation into a 15 features data frame

the deep learning that is studied in this chapter, 15 past values are used for the feature generation.

Once the preparation process has been completed, the next step in a deep learning model development is the configuration of the neural network. In order to create a suitable neural network, several different network configurations are prepared. Recently, a special version of the RNN, the LSTM has been providing great results in many engineering fields. However, modeling complex time series events using LSTM is still a challenging task. In order to provide a more accurate network, the LSTM network is combined with additional neural network types.

The time series dataset used for the training network model contains features generated from the same variable. In such conditions, the features have a very strong inter-correlation, often causing overfitting and less reliable results. To avoid overfitting, the features need to be transformed be less inter-correlated, and more independent from each other. There are several techniques to overcome this phenomenon, but one of the most popular is to use autoencoders [33].

Autoencoder Deep Neural Network

Autoencoders are neural networks that can achieve unsupervised learning. Simply said, it uses backpropagation for learning, by setting the target value the same as the input. In other words, it tries to learn features from features, or approximate an identity function. In the neural network context, autoencoders are a set of fully connected layers, that the input and output dimensions of the autoencoder network are the same. Hidden layers always have less neurons than the input/output layer. Figure 4.17 shows an autoencoder neural network used for the energy demand network configuration. As can be seen, the autoencoder is built with fully connected layers, where the first layer starts with the same number of neurons as the input dimension. Then, the neurons in the second layer are reduced by 50%, and the middle-hidden layer is defined with only four neurons. This is called a bottle neck. After the bottle neck, the dimension of the hidden layers is increased, first to 8, and then to 15.

Training Process of LSTM Deep Learning Model

In order to configure and define the DL model for energy demand, the ANNdotNET tool is used. Once the data preparation has been performed, the network config-

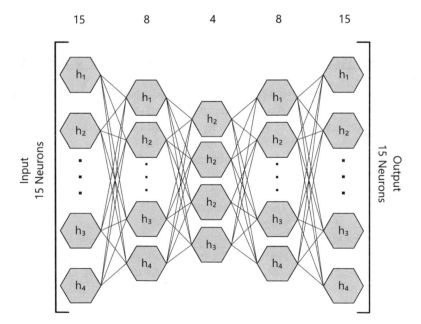

Fig. 4.17 15-8-4-8-15 autoencoder neural network architecture

uration is set up by adding the LSTM layer with a 100 LSTM cell dimension. The LSTM cell is placed after the autoencoder network layer. As a next layer, a DroupOut layer with 30% of dropped values was added. The last layer in the network is the output layer with one neuron. Figure 4.18 shows a schematic deep learning model for energy demand.

The defined model is trained and validated on a 15 features dataset, created from January 2011 to November 2016. The first 80% of the dataset belongs to the training set and the remaining 20% to the validation set. The December 2016 values are used as a test set for comparison analysis between a deep learning model and the TDS model described in the previous section. The training network configuration has been performed using the Adam learner, with the squared error, SE as the loss function, and the root mean square error, RMSE, as the evaluation function. The final model is trained after 5000 epochs.

Fig. 4.18 LSTM deep learning model for energy demand

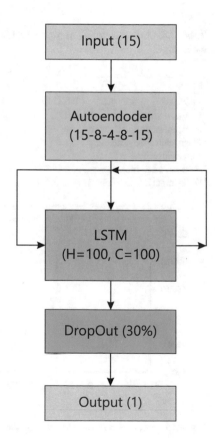

Evaluation Process of the LSTM Deep Learning Model

In order to provide the model evaluation, several performance parameters were calculated against a training and validation set and presented in Table 4.4. A description of each of the performance parameters is given in the literature [15].

From Table 4.4, one can conclude that the model has a high performance for both datasets. The RMSE and R values are roughly equal for both the training and the validation sets. Moreover, the rest of the performance parameters except SE show the same behavior. This is an indication that the model is well trained with a high percentage of accuracy. Figure 4.19 graphically shows the model prediction for the validation set with respect to actual values. The chart series are drawn in different colors (shades) that can be easily seen how prediction values are close to actual values.

Table 4.4 Performance values for the LSTM deep learning model for the training and validation datasets

	SE[a]	RMSE[b]	NSE[c]	PB[d]	R[e]	R[2f]
Train. set	1.20459	0.02648	0.94370	0.035758	0.97363	0.94795
Valid. set	0.27896	0.02550	0.93856	0.03350	0.97096	0.94277

[a]Squared error
[b]Root mean squared error
[c]Nash–Sutcliffe efficiency
[d]Percentage bias
[e]Pearson's coefficient
[f]Determination coefficient

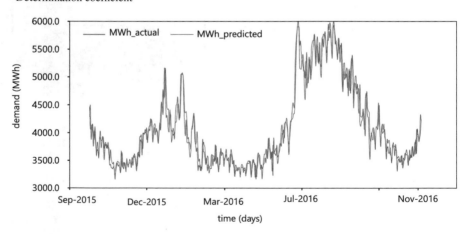

Fig. 4.19 Predicted values calculated by the deep learning model for the validation set

The conclusion that can be established from the model evaluation is that the deep learning model can accurately predict the time period defined by the validation set.

Testing Process of the LSTM Deep Learning Model

Deep learning provides models which can predict future values. Data used for training the deep learning model were not decomposed. In contrast to the TDS model prediction, the data were decomposed, and the trend and random components were included in the modeling. Once the model is calculated, the seasonality component was added, and the prediction (see Fig. 4.15) was calculated. The time series decomposition gives answers for the question as to how data behave in different time periods (seasons), what is the trend of the data in those time periods (seasons)? Deep learning is a black box which trains the model to predict the future values with no additional answers. For this reason, time series decomposition is important in order to prepare and transform the dataset prior to starting deep learning, and also as a comparison analysis between the model predictions.

In order to show how much the deep learning model is accurate, a comparison analysis is performed between the LSTM deep learning and TSD models, using the same dataset. The testing set is represented by the energy demand from December 2016. Table 4.5 summarizes the comparison results.

From Table 4.5 it can be seen that the LSTM deep learning model is significantly better than the TSD model because it is closer to the observed series. However, the comparison chart shown in Fig. 4.20 also shows that the TSD model follows the weekly peaks, but those values are lower than actual values. It can also be noticed that as the prediction period gets longer, the TSD model predicts the values with higher error, while the deep learning model predicts values with a much lower error. The LSTM deep learning model has better RMSE values in all prediction periods: 5 days, 15 days, and monthly. The TSD model has a higher Pearson coefficient, for 5 and 15 days of the prediction, while for the 1 month prediction, the Pearson coefficient is higher for the LSTM deep learning model. The reason why the Pearson coefficient is better for 5 and 15 days for the TDS model may be ignored due to the small dataset, and the RMSE parameter in such a case is more relevant.

Table 4.5 Performance analysis between LSMT and TSD models

	1–5 December 2016		1–15 December 2016		December 2016	
	LSTM	TSD	LSTM	TSD	LSTM	TSD
RMSE	198	353	320	386	281	720
R	0.489	0.923	0.659	0.808	0.829	0.760

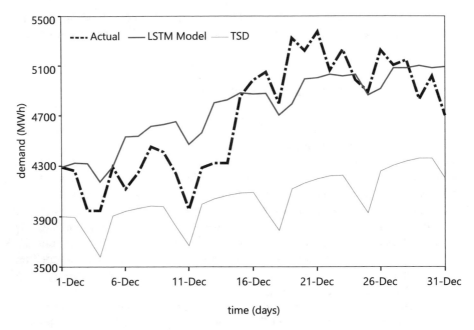

Fig. 4.20 Energy demand prediction for December 2016

Energy demand predictions for December 2016, using the LSTM deep learning model, and the TDS model, with respect to actual values are illustrated in Fig. 4.20. It can be clearly seen that the dashed curve which represents the LSTM deep learning model is much closer to red line, than the TDS model marked with the blue color.

In general, one can say that deep learning model provides a better prediction than the TDS model in all aspects of the analysis.

Deep Learning Model as a Cloud Solution for Smart Cities

In order to prepare, train, and evaluate the deep learning model, the ANNdotNET deep learning tool was developed and used in this study. By using the ANNdotNET, it is possible to incorporate ML tasks into a cloud solution, so that the complete ML process can be automatized and defined into one workflow using cloud services. Once the ML task is incorporated as a cloud solution, it can be part of the bigger smart cities project. In this section details of the possible cloud solution are presented.

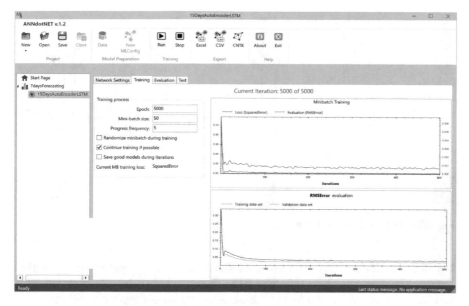

Fig. 4.21 Training module of ANNdotNET deep learning tool

It is very common that an ML solution is split into three phases:

- Data preparation
- Training ML model
- Model Deployment

The first phase consists of a set of data related tasks responsible for the transformation of raw data into a machine ready (mlready) dataset. This phase may include data transformation, outliers identification, features selection, features engineering, cross validation analysis, etc. Once the data is transformed into an mlready dataset, the next phase starts by defining the input and output layers of the deep neural network that is based on the mlready dataset. The input dimension defines the input dimension for the next layer in the network. The network configuration is initialized by providing the mlconfig file [14] that holds information about the network configuration, learning and training parameters.

Once the model configuration is loaded using the mlconfig file, the training process can be started by defining the number of epochs, or by defining the early stopping criteria. The training process can be monitored by reading the training progress information. The information helps the user to decide the training process converging at the expected speed, or when to stop the training process in order to prevent model overfitting. The training module that shows training history is shown in Fig. 4.21. The model deployment is the last phase of the ML cloud solution, and defines several options that can be used for different scenarios. The most common option is to generate a simple web service that contains the implementation of

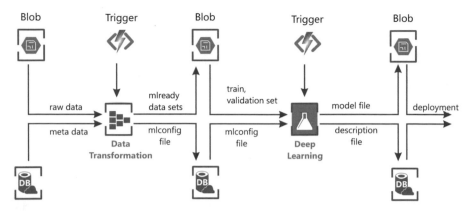

Fig. 4.22 Architecture of an ML cloud solution

the model evaluation. The web service returns the model output in an appropriate format. The model can also be deployed in Excel, to allow the model to behave as an Excel formula. Excel deployment is achieved by implementing additional Excel add-in. The deployment ML model in Excel is usually suitable when dealing with the input data which is relatively easy to represent in Excel.

The complete cloud ML solution is depicted in Fig. 4.22. To implement such a cloud solution, Microsoft Azure Cloud [21] platform can be used. By using the ANNdotNET open source computer program, it is possible to transform data and prepare it for training. Moreover, ANNdotNET provides components for training, evaluation, testing, and deploying deep learning models. Its components can be used in similar cloud solutions depicted in Fig. 4.22, particularly for "data transformation" and "deep learning" cloud solution components.

References

1. Al-Turjman, F., & Alturjman, S. (2018). Context-sensitive access in industrial internet of things (IIoT) healthcare applications. *IEEE Transactions on Industrial Informatics, 14*(6), 2736–2744. https://doi.org/10.1109/TII.2018.2808190
2. Bengio, Y., Simard, P., & Frasconi, P. (1994). Learning long-term dependencies with gradient descent is difficult. *IEEE Transactions on Neural Networks, 5*(2), 157–166. https://doi.org/10.1109/72.279181
3. Bonissone, P. P. (2015). *Springer handbook of computational intelligence.* https://doi.org/10.1007/978-3-662-43505-2
4. Cao, Q., Ewing, B.T., & Thompson, M.A. (2012). Forecasting wind speed with recurrent neural networks. *European Journal of Operational Research, 221*(1), 148–154. https://doi.org/10.1016/j.ejor.2012.02.042
5. Cleveland, R. B., Cleveland, W. S., McRae, J. E., & Terpenning, I. (1990). STL: A seasonal-trend decomposition procedure based on loess. *Journal of Official Statistics.* https://doi.org/citeulike-article-id:1435502

6. Danandeh Mehr, A. (2018). An improved gene expression programming model for streamflow forecasting in intermittent streams. *Journal of Hydrology, 563*, 669–678.
7. Dokumentov, A., & Hyndman, R. J. (2015). STR: A seasonal-trend decomposition procedure based on regression, Department of Econometrics and Business Statistics, Monash University.
8. Gers, F. A., Schraudolph, N. N., & Schmidhuber, J. (2002). Learning precise timing with LSTM recurrent networks. *Journal of Machine Learning Research, 3*(1), 115–143. https://doi.org/10.1162/153244303768966139
9. Graves, A., Mohamed, A., & Hinton, G. (2013). Speech recognition with deep recurrent neural networks. In *IEEE International Conference on Acoustics, Speech and Signal Processing* (Vol. 3, pp. 6645–6649). https://doi.org/10.1109/ICASSP.2013.6638947
10. Hermans, M., & Schrauwen, B. (2013). Training and analyzing deep recurrent neural networks. *NIPS* 2013.
11. Hochreiter, S., & Schmidhuber, J. (1997). Long short-term memory. *Neural Computation, 9*(8), 1735–1780. https://doi.org/10.1162/neco.1997.9.8.1735
12. Hopfield, J. J. (1982). Neural networks and physical systems with emergent collective computational abilities. *Proceedings of the National Academy of Sciences, 79*(8), 2554–2558. https://doi.org/10.1073/pnas.79.8.2554
13. Hoptroff, R. G. (1993). The principles and practice of time series forecasting and business modelling using neural nets. *Neural Computing Applications, 1*(1), 59–66. https://doi.org/10.1007/BF01411375
14. Hrnjica, B. (2018). ANNdotNET- deep learning tool on .Net platform https://doi.org/10.5281/ZENODO.1756095
15. Hrnjica, B., & Danandeh Mehr, A. (2018). Optimized genetic programming applications. IGI Global. https://doi.org/10.4018/978-1-5225-6005-0
16. Hyndman, R. J., & Athanasopoulos, G. (2018). *Forecasting: Principles and practice* (2nd ed.). Melbourne: OTexts. http://OTexts.com/fpp2. Accessed 1 Feb 2019.
17. Kaastra, I., & Boyd, M. (1996). Designing a neural network for forecasting financial and economic time series. *Neurocomputing, 10*(3), 215–236. https://doi.org/10.1016/0925-2312(95)00039-9
18. Lee, T. L. (2008). Back-propagation neural network for the prediction of the short-term storm surge in Taichung harbor, Taiwan. *Engineering Applications of Artificial Intelligence, 21*(1), 63–72. https://doi.org/10.1016/j.engappai.2007.03.002
19. Lu, Y., & Salem, F. M. (2017). Simplified gating in long short-term memory (LSTM) recurrent neural networks. CoRR, abs/1701.0, 5. https://doi.org/10.1109/MWSCAS.2017.8053244
20. Mehrotra, K., Mohan, C. K., & Ranka, S. (1997). *Elements of artificial neural networks*, A Bradford Book (The MIT Press, Cambridge)
21. Microsoft. (2015). Microsoft azure. https://doi.org/10.1007/978-1-4842-1043-7
22. Muhammad, K., Ahmad, J., Lv, Z., Bellavista, P., Yang, P., & Baik, S. W. (2018). Efficient deep CNN-based fire detection and localization in video surveillance applications. *IEEE Transactions on Systems, Man, and Cybernetics: Systems*, 1–16. https://doi.org/10.1109/TSMC.2018.2830099
23. Muhammad, K., Hussain, T., & Baik, S. W. (2018). Efficient CNN based summarization of surveillance videos for resource-constrained devices. *Pattern Recognition Letters*. https://doi.org/10.1016/J.PATREC.2018.08.003
24. Pineda, F. J. (1987). Generalization of back-propagation to recurrent neural networks. *Physical Review Letters, 59*(19), 2229–2232. https://doi.org/10.1103/PhysRevLett.59.2229
25. Rodriguez, C. P., & Anders, G. J. (2004). Energy price forecasting in the Ontario competitive power system market. *IEEE Transactions on Power Systems, 19*(1), 366–374. https://doi.org/10.1109/TPWRS.2003.821470
26. Rosenblatt, F. (1960). Perceptron simulation experiments. *Proceedings of the IRE, 48*(3), 301–309. https://doi.org/10.1109/JRPROC.1960.287598
27. Rumelhart, D. E., Hinton, G. E., & Williams, R. J. (1986). Learning representations by back-propagating errors. *Nature, 323*(6088), 533–536. https://doi.org/10.1038/323533a0

28. Russell, S., & Norvig, P. (2015). *Artificial intelligence a modern approach* (3rd edn.). London: Pearson Education.
29. Sak, H., Senior, A., & Beaufays, F. (2014). Long short-term memory recurrent neural network architectures for large scale acoustic modeling. Interspeech 2014, (September), pp. 338–342. https://doi.org/arXiv:1402.1128
30. Schmidhuber, J. (2015). Deep learning in neural networks: An overview. *Neural Networks, 61*, 85–117. https://doi.org/10.1016/j.neunet.2014.09.003
31. Taylor, S. J., & Letham, B. (2018). Forecasting at Scale. *American Statistician, 72*, 37–45. https://doi.org/10.1080/00031305.2017.1380080
32. Ullah, A., Muhammad, K., Del Ser, J., Baik, S. W., & Albuquerque, V. (2018). Activity recognition using temporal optical flow convolutional features and multi-layer LSTM. *IEEE Transactions on Industrial Electronics*. https://doi.org/10.1109/TIE.2018.2881943
33. Vincent, P., Larochelle, H., Lajoie, I., Bengio, Y., & Manzagol, P.-A. (2010). Stacked denoising autoencoders: Learning useful representations in a deep network with a local denoising criterion. *Journal of Machine Learning Research*. https://doi.org/10.1111/1467-8535.00290
34. Yu, D., Eversole, A., Seltzer, M., Yao, K., Kuchaiev, O., Zhang, et al. (2014). An introduction to computational networks and the computational network toolkit. Microsoft Research.

Chapter 5
RETRACTED CHAPTER:
Context-Aware Location
Recommendations for Smart Cities

Akanksha Pal and Abhishek Singh Rathore

Abbreviations

IoT   Internet of Things
LBSN   Location-Based Social Networks
ML   Machine Learning

Introduction

Smart cities are due for a technical upgrade. With the help of machine learning (ML), for example, a smart city's system can tell waste haulers when garbage cans need to be emptied and smart buildings can alert maintenance staff of impending repair requirements. In a smart city, ML can greatly change the morning commute by analyzing real-time traffic and managing traffic lights with the help of cameras and sensors, thus reducing travel time and congestion.

In agriculture, a smart city's network provides optimized information and an expanded computing capacity with millions of inputs. With the help of ML, recent prototypes are able to autonomously navigate a farm to identify all plant types and remove only the weeds. ML also has shortened the distance between cities and farms. Urban planners and farmers are better able to make the right decision for planting dates with minimum resources and energy.

The original version of this chapter was retracted: The retraction note to this chapter is available at https://doi.org/10.1007/978-3-030-14718-1_12

A. Pal (✉) · A. S. Rathore
Shri Vaishnav Institute of Information Technology SVVV, Indore, India
e-mail: er.akankshapal@gmail.com

© Springer Nature Switzerland AG 2020
F. Al-Turjman (ed.), *Smart Cities Performability, Cognition, & Security*,
EAI/Springer Innovations in Communication and Computing,
https://doi.org/10.1007/978-3-030-14718-1_5

Like ML, recommendation systems also play an important role in smart cities. The smart city is a product of advanced development in the new era of information technology, where technological solutions are being developed to deal with the challenges of the city. Location recommendations are a vital part of making cities smart. Users may not always be decisive about choosing a location to visit. Thus, a location recommendation program that is able to observe the user's preferences can suggest a nearby location. For example, applications can be developed to suggest dining establishments based on a user's location, past experiences, ratings, and reviews. Similarly, a route recommendation program can provide travel time and a fuel-efficient route based on a user's live location and destination [1].

In their daily lives, people depend on the recommendations of family, friends, media, and social networks, among others. Recommendations are also provided by many e-commerce websites, especially location recommendations. The recommendation system can provide a variety of locations in which it believes the user will be interested. This system saves the user time by providing quick information and recommendations by knowing the user's interests and favorites [2, 3].

Location recommendation systems suggest places, routes, and events to users based on two primary considerations: 1) spatial data (e.g., the user's current location, travel route, nearest prominent sites) and 2) the user's preferences and past history of visits. An algorithm then computes the relevance of nearby objects and suggests the ones with the highest probabilities.

Location-based social networks (LBSN) such as Facebook and Twitter, also track users' locations and preferences. In addition, marketing companies often provide this service to their users. These systems can filter information based on searches, social links, previous check-ins, other customer ratings, and previous ratings given by the user. LBSNs also use this information to provide recommendations for places and events to other connected users. Google Places is a popular example of a recommendation system that offers its users relevant suggestions for nearby locations they can visit.

One of the challenges faced by recommendation systems and similar big data applications is the various amounts of data that have to be filtered and which data sets or sources to use. These systems can be studied with respect to classifications based on the data source, the method or algorithm selected (content-based, link-based, or collaborative methods), and the purpose of the recommendation. Other studies also considered sentiments (e.g., trust, friendship ties) that influence a user's acceptance of recommendations in order to develop more accurate systems. The outcomes—how much these recommendations actually influence a user's experiences and decisions—are also worth analyzing.

The Internet of Things (IoT) is defined as a group that contains users, computing devices, and objects that are interlinked or related to each other. These objects can transfer data over the internet without the need for any human or computer interaction. Smart televisions and smart security systems are some examples of IoTs. For example, when a smart security system detects an intruder, it will automatically inform the owners and send a message to the local police, without requiring any permission or instructions from a human. IoTs are widely used in smart city applications, medical applications, security, industry, and homes. An IoT-based medical systems may be one of the most important application. Such a system makes use of the 5G technology in newer medical devices, which can react and interact with different sensors through safe wireless medical sensor networks.

IoT is widely used in cloud-based applications to provide a secure application that can avoid faulty data. It also plays an important role in creating smart homes. People with disabilities and elderly people can use this technology to provide smart assistance, such as voice control that is connected to a hearing aid. This technology also can assist blind persons with their surroundings and objects around them. In addition, it plays an important role in industrial applications by operating process controls and operator tools, as well as regularly checking the safety and security of the industry through the data from the information systems.

Existing Frameworks

In this section, we briefly discuss previously published work in this field. Wang et al. [3] discussed ML and data mining as tools for recommendation systems. Bian and others [4] attempted to implement a collaborative filtering method that was focused on the interests and behavior of customers. Linden [4] took the user's past history of purchases and businesses, as well as knowledge of their likes and patterns, to recommend new products or places.

Other algorithms that have been discussed in research publications are the K-mean algorithm, which was validated for social networking websites [5] and matching graphs. G.L.P. [6] designed recommendation systems by finding a measure of similarity after applying the k-mean algorithm. Kwon [7] described an innovative method that was a cross between collaborative filtering and cluster methodologies.

Yan Zhao [8] analyzed a user's location data along with contacts in a location recommendation application to suggest friends. Liangbin Yang [9] used contact data via the filtering method to design a more efficient procedure. A hybrid was introduced by Zhoubi Deng [10] for big datasets, which was a cross between content-based and COF algorithms.

Fei Yu solved a problem related to the large amount of user data available by using a greedy algorithm [11]. Zhen Zhang analyzed recommendations using a binary classification algorithm, whereas Zhou Zhang solved recommendation systems as problems of total probability and FRUTAI [12, 13]. An even more complex approach by L.T. Nguyen inferred a user's opinion or inclination about friends in order to give friend suggestions on social networking sites [14].

A unique approach was taken by Linke Guo in which user information, interests, and traits were used to find friends among strangers connected by a multiple-node chain of friends [15]. Fizza Abbas [16] based recommendation systems on trust factors and the relation of trust with the probability of users accepting a recommendation. She proposed a recommendation system based on computing a measure of trust by looking at friend chains and reliability based on the social platform. Privacy and security are also needed for smart cities [17, 18].

Another unique solution was given by Pankaj L. Pingate [19], who developed a recommendation system based on similar interests and lifestyles instead of social relations. Similar work regarding ML-based smart cities has been published in various papers [20].

Proposed Work

Context-awareness systems have been a growing research field, especially for recommendations, information recovery, and auto-adoptive applications. The basic element in these emerging standards is the concept of context [21]. The concept of context is presented through a context metamodel and a proposed architecture of such a model in Figs. 5.1 and 5.2.

In the model, context is defined as a combination of different forward-facing groups of properties: the channel, location, and time description of the user and the system. System activity describes the system states and a user profile to explain the user information in detail. The channel defines the medium by which the information of the context can be transmitted. The system activity describes the system states, such as if it is on standby or in any other state. The location and time descriptions identify the position of the user—that is, where the user is located while interacting with the application. The user profile is structured, organized, maintained, and collected information. Many similar models have also been proposed to make cities smarter, especially with recommendation systems using ML algorithms.

A newly proposed approach for a trust-based location recommendation [22] depends on the following five main measurement stages:

1. Weight measurement
2. Edge trust measurement
3. Node trust measurement
4. Edge influence measurement
5. Node influence measurement

The approach is applied to the directed graphs—that is, $G (V, E)$, where V is a set of nodes representing friends in a social network and E is a set of edges representing relationships between those friends. Each edge is assigned a value, known as a weight value, to describe the related attributes of friends. Although the friendship strength weight between two nodes, f_i and f_j, is constant, their measured values differ because of unbalanced friendship relations (Fig. 5.3).

The proposed approach presented in the above algorithm follows the following procedure: for each f_i in G, let the weight represent the calculated correlation value between f_i and f_j. In the following step, this weight is used to calculate the edge trust, which represents the trust f_i has in f_j. Later, the node trust is calculated as the measurement that trust f_i has on the whole network [22]. Depending on the calculated edge trust and considering the number of common visited locations

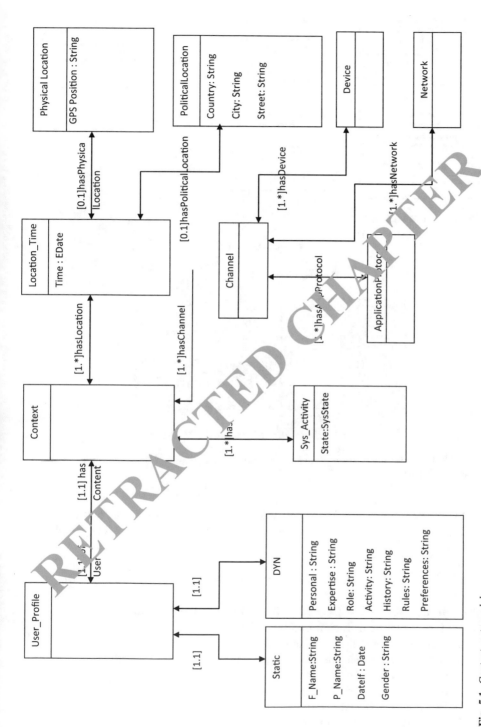

Fig. 5.1 Context metamodel

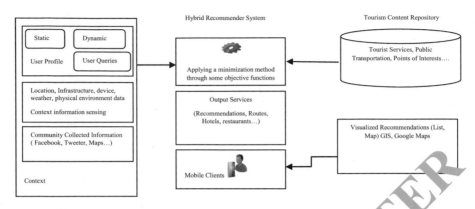

Fig. 5.2 Proposed architecture

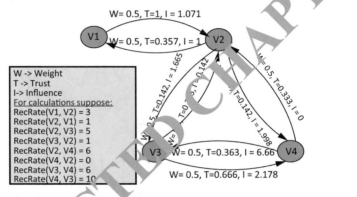

Source File - https://ieeexplore.ieee.org/document/8265107

Fig. 5.3 Experimental friendship strength between nodes. Source: https://ieeexplore.ieee.org/document/8265107

between every two friends in the network, another parameter is measured: the edge influence, which represents the impact of one friend on the second. The weight is calculated as follows:

$$\text{Weight}\left(f_{ri},\, f_{rj}\right) = \text{cor}\left(\text{fr}_{i\text{attr}}, \text{fr}_{j\text{attr}}\right)$$

Algorithm 1 Trust-Based Location Recommendation Algorithm
1: vec1 ← vector holds each friend ID and weight
   2: res1 ← vector holds the weight value for each friend in vec1
   3: vec2 ← vector holds the row ID for each two connected friends
   4: res2 ← vector holds the edge trust for each friend in vec2
   5: res3 ← vector holds the edge influence for each node in vec2
   6: for all *friends* in Dataset do
   7: Weight(*fi*, *fj*) = cor (*fiattr*, *fjattr*)

8: end for
9: for all *friends* in Dataset do
10: vec1 = Dataset[rowID, (“*fi*”,”Weight”)]
11: res1 = Dataset[which(“*fi*” == vec1[1]), ”Weight”]
12: Dataset[rowID, “EdgeTrust ”] = vec1[2]/sum(res)
13: end for
14: for all *friends* in Dataset do
15: vec2 = Dataset[rowID, c(“*fi*”,”*ff*”)]
16: res2 = Dataset[which(“*ff*”==vec2[1]),”EdgeTrust ”]
17: Dataset[rowID, “NodeTrust”] = sum(res2)
18: for all *friends* in Dataset do
19: vec2 = (Dataset[rowID, c(“*fi*”, “*ff*”)])
20: counterpart = Dataset[which(“*fi*”=vec2[2] and
“*ff*”=vec2[1]), “EdgeTrust”]
21: Dataset[rowID,“EdgeInfl”] = Dataset[rowID,“RecRate”]* ounterpart
22: end for
23: for all *friends* in Dataset do
24: vec2 = Dataset[rowID,c(“*fi*”,“*ff*”)]
25: res3 = Dataset[which (“*fi*” = vec2[1]), “Edge fl”]
26: Dataset[rowID,“NodeInfl”] = sum(res3) * Dataset[rowID,“NodeTrust”]
27: end for

Trust is an important aspect in defining the strength and deepness of a relationship. A highly trusted friend's recommendation seems likely to be accepted by the other. There are two types of trust in the above proposal: edge trust and node trust.

Edge trust is the trust value each friend has in the other. If the data set has m values, then the edge trust can be calculated as follows:

$$\text{EdgeTrust}\left(f_{ri}, f_{rj}\right) = \text{Weight}\left(f_{ri}, f_{rj}\right) / \sum_{x=1}^{m} \text{Weight}\left(f_{ri}, f_{rx}\right)$$

Node trust is the measure of trust a friend has in a social network. This can be calculated as the sum of all the trust values that other friends have in the subject friend. Simply, it can be written as follows:

$$\text{NodeTrust}\left(f_{ri}\right) = \sum_{x=1}^{m} \text{EdgeTrust}\left(f_{rx}, f_{ri}\right)$$

In smart cities, such algorithms are helpful because users tend to visit places recommended by the recommendation applications, then leave feedback on the applications, which helps to improve the recommendation and hence increases its trust among the users. This could be useful for both place recommendations and path recommendations. Path-recommending applications can use different algorithms to measure the shortest path depending upon the user's live location and desired destination [1]. There are many options depending upon the different complexities and the use of weighted and unweighted graphs. Some of more commonly used approaches are Dijkstra's algorithm, the Bellman-Ford algorithm, and breadth-first search.

Discussion

Popular sites such as Rotten Tomatoes and YouTube, which has by far the largest community, use recommendation systems for users. These sites employ a user's history of watched or searched videos to provide future suggestions. Facebook takes a user's friend links to provide friend suggestions. However, these are not always the most accurate approaches. Sites such as Amazon and Flipkart analyze and compare personality traits for better results. In this case, when a person logs into a website, the person's search history is obtained by cookies. The history is studied and the data are fed into the recommendation systems of the website, which suggests several options that the user may like. In addition, products that a user searches for on shopping websites can be shown in advertisements while the user is viewing Facebook or Instagram. This is also done with the help of recommendation systems.

The method proposed in this chapter is an important feature for smart city development. Using recommendation systems based on a ML model will provide accurate and fault-free methods compared to other available recommendation systems. These types of recommendation systems can be used in medical applications, activity-tracking watches that provide helpful suggestions for the person's fitness based on history, and security systems. They can also be used in devices for disabled people that record and study a person's daily routines, allowing the system to provide better assistance for them.

Smart televisions automatically suggest programs and display these suggestions based on their recommendation systems. Recommendation systems are also used in transportation systems; for example, when a person starts a car and enters his or her location, the system automatically collects the user's previous history and suggests a number of routes for travelling. If the person had used inexpensive modes of transportation in the past, then it also could show various ways to reach the destination at low cost.

These examples mostly make use of location recommendation systems, which contain data obtained from a person's history. When these data are fed into the machine model, it produces results that are suggested to the user.

Conclusion

In this chapter, we demonstrated the importance of location recommendation systems for smart cities, which are based on IoT using machine learning. Although recommendation systems are being studied extensively, few approaches consider that the proposed algorithms might have more trust and influence on network members. In this work, we suggest that the following components of a location recommendation system are important to users: public feedback, live location, frequency of visits to a specified place, and the level of trust in one place. These algorithms can be updated accordingly based on the results of future studies.

References

1. Guo, L., Zhang, C., & Fang, Y. (2013). *A trust-based privacy-preserving friend recommendation scheme for online social networks*. Piscataway, NJ: IEEE.
2. Kacchi, T. R., & Deoranker, A. V. (2016). Friend recommendation system based on lifestyles of users. In *International Conference on Advances in Electrical, Electronics, Information, Communication and Bioinformatics (AEEICB16)* (pp. 01–04).
3. Wang, P. Z., & Qi, H. (2015). Friendbook: A semantic-based friend recommendation system for social networks. *IEEE Transactions on Mobile Computing, 14*(3), 1–14.
4. Bian, L., & Holtzman, H. (2011). Online friend recommendation through personality matching and collaborative filtering. In *Proceedings of the 5th International Conference on Mobile Ubiquitous Compute, Systems, Services and Technologies* (pp. 230–235).
5. Amazon. (2014). [Online]. Available: http://www.amazon.com
6. Netflix. (2014). [Online]. Available: https://signup.netflix.com/
7. Kwon, J., & Kim, S. (2010). Friend recommendation method using physical and social context. *International Journal of Computer Science and Network Security, 10*(11), 116–120.
8. Du, Z., Hu, L., Fu, X., & Liu, Y. (2014). *Scalable and explainable friend recommendation in campus social network system*. Dordrecht: Springer.
9. Yang, L., Li, B., Zhou, X., & Kang, Y. (2018). *Micro-blog friend recommendation algorithms based on content and social relations*. Singapore: Springer.
10. Zhao, Y., Zhu, J., Jia, M., Yang, W., & Zheng, K. (2017). *A novel hybrid friend recommendation framework for Twitter*. Cham: Springer.
11. Raghuwanshi, R., & Prajapati, G. L. (2017). An approach for friend recommendation based on selected attributes. In *Proceedings of the World Congress on Engineering 2017*, Vol. II WCE 2017, July 5–7, 2017, London, UK.
12. Deng, Z., He, B., Yu, C., & Chen, Y. (2012). *Personalized friend recommendation in social network based on clustering method*. Berlin/Heidelberg: Springer.
13. Farikha, M., Chihi, M., & Gargouri, F. An user interest ontology based on trusted friend preferences for personalized recommendation. In *EMCIS 2017, LNBIP 299* (pp. 54–67). Cham: Springer.
14. Fei, Y., Che, N., Li, Z., Li, K., & Jiang, S. Friend recommendation considering preference coverage in location-based social networks. In *PAKDD 2017, Part II, LNAI 10235* (pp. 91–105). Cham: Springer.
15. Zhang, Z., Zhao, X., & Wang, G. (2017). *FE-ELM: A new friend recommendation model with extreme learning machine*. New York: Springer.
16. Zhang, Z., Liu, Y., Ding, W., & Huang, W. W. (2015). A friend recommendation system using users' information of total attributes. In *ICDS 2015, LNCS 9208* (pp. 34–41). Cham: Springer.
17. AI-Turjman, F., & AITurjman, S. (2018). Confidential smart-sensing framework in the IoT era. *The Springer Journal of Supercomputing, 74*(10), 5187–5198.
18. Alabady, S. A., AI-Turjman, F., & Din, S. (2018). A novel security model for cooperative virtual networks in the IoT era. *Springer International Journal of Parallel Programming*, 1–16.
19. Nguyen, T. L. T., & Cao, T. H. (2014). Multi-group-based user perceptions for friend recommendation in social networks. In *PAKDD 2014, LNAI 8643* (pp. 525–534). Cham: Springer.

20. AI-Turjman, F., & AITurjman, S. (2018). Context-sensitive access in industrial internet of things (IIoT) healthcare applications. *IEEE Transactions on Industrial Informatics, 14*, 2736–2744.
21. Kumar, P., & Reddy, G. R. M. (2018). *Friendship recommendation system using topological structure of social network*. Singapore: Springer.
22. https://ieeexplore.ieee.org/document/8265107

Chapter 6
Fractional Derivatives for Edge Detection: Application to Road Obstacles

Roy Abi Zeid Daou, Fabio El Samarani, Charles Yaacoub, and Xavier Moreau

Introduction

Sensor networks and the Internet of Things (IoT) [1] are key technologies towards the development of smart cities. While sensing, processing, and communicating data have long been studied, adapting conventional technologies to modern contexts and requirements such as large data volume, high data mobility, and real-time processing is a challenging issue. Different challenges in realizing intelligence in smart cities have been addressed in [2]. Optimizing data delivery within limited resources in highly dynamic topologies is investigated and proposed in [3] and a platform for securing the delivery process of sensed data is presented in [4].

On the other hand, the last decade has witnessed a rise in autonomous vehicles technology, driving towards a paradigm shift in transportation systems [5]. Since the development of vehicle intelligence is key in reducing congestion, pollution and

R. Abi Zeid Daou (✉)
Faculty of Public Health, Biomedical Technologies Department, Lebanese German University, Jounieh, Lebanon

MART Learning, Education and Research Center, Chnaniir, Lebanon
e-mail: r.abizeiddaou@lgu.edu.lb; roydaou@mart-ler.org

F. El Samarani
Faculty of Engineering, Holy Spirit University of Kaslik, Jounieh, Lebanon

IMS Laboratory, Group CRONE, Bordeaux University, Bordeaux, France

C. Yaacoub
Faculty of Engineering, Holy Spirit University of Kaslik, Jounieh, Lebanon
e-mail: charlesyaacoub@usek.edu.lb

X. Moreau
IMS Laboratory, Group CRONE, Bordeaux University, Bordeaux, France
e-mail: xavier.moreau@u-bordeaux.fr

© Springer Nature Switzerland AG 2020   115
F. Al-Turjman (ed.), *Smart Cities Performability, Cognition, & Security*,
EAI/Springer Innovations in Communication and Computing,
https://doi.org/10.1007/978-3-030-14718-1_6

accidents, while at the same time, improving mobility, we focus in this chapter on detecting road obstacles for driverless control of vehicles in the context of smart transportation and safe cities.

The use of powerful computational software and fast hardware processors has made image processing a main ingredient in several control systems due to the precision and low cost it offers compared to other similar sensing systems.

On the other hand, fractional calculus has recently emerged in various engineering fields due to the good results and robustness such computation has offered to different fields, such as electrical engineering [6, 7], automotive engineering [8, 9], heat diffusive interfaces [10–12], signal processing and filtering [13–15], and many others [16].

Although fractional calculus is a very old mathematical idea that was raised in letters exchanged between L'Hopital and Leibniz in 1695 [17, 18], the first application in this domain remained absent until 1975 when Oustaloup implemented a controller of order 3/2 in order to direct a laser beam [19].

Thus, the main objective of this work is to implement this calculation technique for edge detection. This technique will be applied on some gradient-based edge detection methods. Integer order and fractional order approaches will be applied to images containing road obstacles. We have chosen to treat such images as the obstacles found on roads are of major concern and can lead to car accidents resulting in severe injuries and deaths, especially when driving at high speeds. Thus, the automatic identification of these obstacles would gain a great deal of importance if exploited in the context of autonomous driverless vehicles and safe cities.

Three different types of road obstacles have been identified so far: speed bumps, speed humps, and speed cushions. Other road uncertainties have also been classified as road cracks and potholes; however, these latter are beyond the scope of this chapter and will be treated in a future work. The fractional order and the integer order detection methods, based this time on Sobel gradient technique, will be implemented in order to detect the first three already listed obstacles, that is, bumps, humps, and cushions. Both methods will be compared in terms of edge detection performance and computational load.

No similar research has been conducted so far. In fact, fractional order filters have been used for specific operations such as feature extraction and segmentation [20–22]. As for road obstacles detection, no remarkable works have been identified, mainly based on conventional imaging. However, some research has been published in this domain based on different sensing tools as laser scanners, lidar, and radars [23–26].

Hence, this chapter presents a novelty in the way of detection of road abnormalities using fractional calculus techniques. In addition, the classification of these abnormalities was not also found in the state of the art.

Thus, this chapter is organized as follows: in Sect. 6.2, an overview of the fractional calculus and its applications in the different engineering domains is be presented. In addition, some mathematical definitions for the integration and derivative are also mentioned. Section 6.3 presents the edge detection techniques. At this stage, the integer-order methods are explored. They are classified based on gradient derivation techniques (the Laplacian techniques are not treated in this chapter). In addition to that, the computations of the fractional order filters are also presented at this level. In Sect. 6.4, the edge detection techniques are applied to identify road abnormalities, mainly humps, bumps, and cushions. Results are discussed in Sect. 6.5, whereas the applications that can exploit these results are proposed in Sect. . Finally, Sect. 6.7 summarizes the achieved work and proposes some ideas which may enrich the whole work.

Overview of the Fractional Calculus

As already mentioned in the introduction, the idea of fractional calculus is very old and dates back to three centuries. This idea was based on some letters exchanged between two well-known mathematicians at that era, L'Hopital and Leibniz, concerning the significance of this derivative and how it would be implemented in a real system. The answer to that question came after 280 years when the first application concerning this idea was implemented. It was physically embedded within a controller of order 3/2.

Since then, the applications involving fractional calculus have increased dramatically due to the success such systems have shown compared to the integer order systems. These applications have treated process identification, modeling, and control [27].

Concerning the identification and the modeling parts, it has been shown that many transfer functions of systems are modeled using fractional order components as the heat diffusive phenomenon in homogeneous metals [28, 29], viscous–thermal losses in pipelines [30], chaotic systems [31, 32], expiration and transpiration of the lungs [33], muscle contraction and relaxation [34], and drug effects [35].

As for the control, lots of fractional order controllers have been developed. The first one was the CRONE (French acronym of "Commande Robuste d'Ordre Non Entier," which signifies non-integer order controller) controller [36]. Then, an upgraded version of the PID has shown the integration of fractional order integrator and derivative to get the $PI^\lambda D^\mu$ [37]. Other controllers have also shown a shifting towards the fractional calculus as the sliding mode control (SMD) [38], the adaptive controllers [39], and the observers [40].

As for the physical implementation of these controllers, Oustaloup has shown in 1995 an innovative way to present a fractional order integration or derivative using a series of resistors (R) and capacitors (C) in two different arrangements: gamma RC arrangement and parallel arrangement of RC elements placed in series. The values of the components are computed after defining some recursive and real poles and zeros which represent the rationalized transfer function. Note that four cells are enough to represent any fractional order system. Interested readers can refer to the following references for more details: [41–43].

Recently, fractional calculus has emerged in new engineering domains as the modelling of earthquake and seismic problems [44, 45], the economical market model [46, 47], and many others.

As for the research in this field, the increasing number of books, journals, conferences, and publications discussing fractional calculus and/or applications has shown a global interest in the research community, mainly due to the results this mathematical tool has been delivering in lots of domains and the numerical and analytical methods found to help such computations [48].

Going back to the core of this chapter, fractional calculus has been recently applied to image processing. Few published works present the procedure and compare the results between the integer-order and the fractional-order methods. Before 2010, most papers treating this research topic focused on the use of fractional Fourier transform for compression and encryption [49, 50]. After that, the fractional calculus tool was used for denoising [51, 52] and segmentation [53].

Concerning edge detection, few studies have implemented fractional tools trying to improve the already existent techniques. In 2003, Mathieu et al. presented a fractional differentiation method for edge detection without relying on gradient and Laplacian techniques. The obtained results showed that the edges are getting thinner. However, the proposed method was only applied over some mathematical functions as 2D and 3D parabolic forms [54]. In 2010, Yang et al. proposed a new edge detection operator. The proposed YE&YANG operator was compared to Canny and CRONE operators and it has been shown that this new technique eliminates the smoothing preprocessing and provides a new approach to tune the compromise between noise immunity and detection accuracy [55]. Nowadays, several researchers work in this domain, mainly ElAraby et al. [56], Hacini et al. [21], and Wang et al. [23, 24].

As for the relevant applications, most works related to image processing, in particular to edge detection, are applied to medical domain for feature extraction as the case of MRI brain image processing [57], cardiac ultrasound image [58], and Alzheimer disease on MRI images [59].

Concerning obstacle avoidance, some works have been achieved using fractional calculus applied for the classification [60] or for the control part of the system [61]. Thus, so far, road obstacles have not been approached using edge detection techniques, more precisely using fractional edge detection tools; this idea will constitute the originality of this work.

Edge Detection Techniques

In image processing, an edge is an intentionally abrupt change in intensity, in contrary to noise (undesirable effect) [62]. This definition leads to the use of derivations in order to localize edges in an image. Edge detection methods are many. Actually, spatial edge detection techniques can be grouped into two main categories: gradient and Laplacian.

The first method is based on the use of first order derivative in order to define the edge whereas the second method relies on the use of two successive derivations (i.e., second order derivative) in order to find the edge. Each one has its advantages and limitations and several techniques have been developed while using these mathematical operations.

Gradient-based methods rely on the first derivative of an image in order to detect its edges. Some of the algorithms based on the gradient are Roberts, Prewitt, and Sobel [63, 64].

The Laplacian uses the zero crossings in the second derivative of an image. Some of the methods based on the Laplacian method are Marr–Hildreth, LoG (Laplacian of Gaussian), and Canny [65].

Later on, we will concentrate on the first approach only as the fractional computation will be applied to the three already listed methods (Roberts, Prewitt, and Sobel). As for the Laplacian fractional methods, they will be presented in a future work.

Conventional Methods

The algorithm for gradient methods is shown in Fig. 6.1. In this flowchart, one can notice that the input image is first converted to an intensity image (or gray scale image), before its convolution with the vertical and horizontal masks, each apart. The last phase consists of summing up the absolute value of the results containing, respectively, the vertical and the horizontal edges to get the output image with edge pixels obtained after a thresholding operation.

Concerning Roberts method, it was introduced in 1965. It executes, on an image, an easy and fast 2D spatial gradient calculation. This process sheds light on high

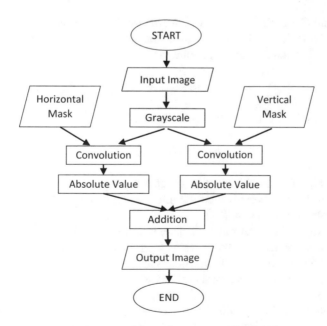

Fig. 6.1 Flowchart of the gradient methods (Roberts, Prewitt, and Sobel)

Table 6.1 Roberts masks

+1	0	0	+1
0	−1	−1	0
I_x		I_y	

Table 6.2 Prewitt masks

−1	−1	−1	−1	0	+1
0	0	0	−1	0	+1
+1	+1	+1	−1	0	+1
I_x			I_y		

spatial frequency areas, including edges. Each value of the output pixels at a precise point corresponds to the absolute estimated magnitude of the gradient at this point of the input image [66]. In this method, Roberts uses two masks as shown in Table 6.1. One is similar to the other but rotated by 90°.

These masks are arranged to mark, in the best way, the edges located at 45° to the grid of pixels; one mask for each of the axes. These masks can covert the image separately and give two separate measurements, each for an orientation (I_x for vertical and I_y for horizontal directions). The next step consists of the combination of the two measurements in order to define the absolute magnitude of the gradient at every point to define whether the pixel is an edge or not. Thus, some calculations have to be made. The magnitude is given by:

$$|G| = \sqrt{G_x^2 + G_y^2}, \tag{6.1}$$

where G_x and G_y are the result of filtering the image with I_x and I_y, respectively.

In order to facilitate the computation and make it faster, we can simplify the above expression and estimate it as follows:

$$|G| = |G_x| + |G_y|. \tag{6.2}$$

Also, the angle of orientation is estimated as follows:

$$\theta = \tan^{-1}\left(G_y/G_x\right) - 3\pi/4. \tag{6.3}$$

In general, orientations are not as useful as gradient magnitude for edge detection, but they do complement the information extracted from an image using the magnitude of the gradient [67].

Thus, the algorithm determines if a pixel is an edge or not. In fact, by thresholding the result, pixels where we have a remarkable change in intensity are marked as edges, while the remaining non-edge pixels are nulled.

As for Prewitt, this method computes an estimation of the gradient of the intensity function of an image. It is a 2D gradient operator (shown in Table 6.2) applied in both vertical and horizontal directions. However, this approximation is quite basic, particularly when using images with high frequency alternations [68].

Table 6.3 Sobel masks

−1	−2	−1	−1	0	+1
0	0	0	−2	0	+2
+1	+2	+1	−1	0	+1
I_x			I_y		

The masks for this algorithm, shown in Table 6.2, are of a 3×3 size; one mask is similar to the other but rotated by $90°$. They are designed to give better results for edges that run horizontally (I_x) and vertically (I_y) relative to the pixels grid [69].

The magnitude follows the same calculation as Roberts' method but the angle of orientation for Prewitt is different, it is equal to:

$$\theta = \tan^{-1} \left(G_y / G_x \right). \tag{6.4}$$

As for Sobel edge detection method, it was developed in 1970. It finds edges using Sobel approximation to the derivative. It follows the edges on these points where the highest gradient is located [70]. This procedure is performed on a 2D level so it marks the areas with the highest frequency equivalent to edges. It is basically similar to Prewitt's method, where the only two differences are the masks displayed in Table 6.3, and the orientation.

2×2 masks are of simpler conception, but they are not as useful for computing edge direction as masks that are symmetric about the center point; the smallest of which are of size 3×3. The main advantage in this case is by using a mask with an odd size; the operator will be centered, and we will get an estimate based on the center pixel.

Fractional Methods Implementation

Many definitions have been proposed concerning the fractional derivation and integration, from among the ones listed by well-known mathematicians that have worked on this idea, as Riemann–Liouville (R-L), Grümwald–Letnikov (G-L), and Caputo [17]. Each one of them has defined a relation for the fractional order differentiation. Thus, the integration of the fractional calculus consists of calculating new masks, other than the ones shown in Tables 6.1, 6.2, and 6.3, while using the G-L technique. In the following, the new masks will be defined for the already listed methods.

Before computing the new masks for Roberts, Sobel, and Prewitt edge detection methods, let us define the context of our work: having a function $f(x, y)$ and σ being the order of the fractional differential, computing the fractional partial differential equation, based on G-L, can be represented as follows, for the x and y orientations:

$$\frac{\partial^\sigma (f(x, y))}{\partial x^\sigma} = f(x, y) - \sigma f(x - 1, y) - \frac{\sigma(-\sigma + 1)}{2} f(x - 2, y)$$

$$- \frac{\sigma(-\sigma + 1)(-\sigma + 2)}{6} f(x - 3, y),$$

$$(6.5)$$

$$\frac{\partial^\sigma (f(x, y))}{\partial y^\sigma} = f(x, y) - \sigma f(x, y - 1) - \frac{\sigma(-\sigma + 1)}{2} f(x, y - 2)$$

$$- \frac{\sigma(-\sigma + 1)(-\sigma + 2)}{6} f(x, y - 3)$$

$$(6.6)$$

With Roberts method for first order derivative, one can obtain the filter output given the original filter masks as:

$$G_x = f(x - 1, y - 1) - f(x, y), \tag{6.7}$$

and

$$G_y = f(x - 1, y + 1) - f(x, y). \tag{6.8}$$

Applying a fractional order derivation on G_x, with an intensity factor of ½ [57]:

$$G_x{}^\sigma = \frac{1}{2} \left(\frac{\partial^\sigma (f(x - 1, y - 1))}{\partial x^\sigma} - \frac{\partial^\sigma (f(x, y))}{\partial x^\sigma} \right). \tag{6.9}$$

Therefore, applying (6.5) on the two terms in (6.9), one can get:

$$\frac{\partial^\sigma (f(x - 1, y - 1))}{\partial x^\sigma} = f(x - 1, y - 1) - \sigma f(x - 2, y - 1)$$

$$- \frac{\sigma(-\sigma + 1)}{2} f(x - 3, y - 1)$$

$$- \frac{\sigma(-\sigma + 1)(-\sigma + 2)}{6} f(x - 4, y - 1),$$

$$(6.10)$$

and

$$\frac{\partial^\sigma (f(x, y))}{\partial x^\sigma} = -f(x, y) + \sigma f(x - 1, y) + \frac{\sigma(-\sigma + 1)}{2} f(x - 2, y)$$

$$+ \frac{\sigma(-\sigma + 1)(-\sigma + 2)}{6} f(x - 3, y). \tag{6.11}$$

Given (6.10) and (6.11), a new mask I_x can be obtained for $G_x{}^\sigma$ as represented in Table 6.4.

Table 6.4 Fractional order Roberts filter mask I_x

Offset : x ↓ y→	-1	0
-4	$-\frac{\sigma(-\sigma+1)(-\sigma+2)}{12}$	0
-3	$-\frac{\sigma(-\sigma+1)}{4}$	$-\frac{\sigma(-\sigma+1)(-\sigma+2)}{12}$
-2	$-\frac{\sigma}{2}$	$-\frac{\sigma(-\sigma+1)}{4}$
-1	$\frac{1}{2}$	$-\frac{\sigma}{2}$
0	0	$\frac{1}{2}$

Table 6.5 Fractional order Roberts filter mask I_y

Offset : x ↓ y→	-3	-2	-1	0	1
-1	0	$-\frac{\sigma(-\sigma+1)(-\sigma+2)}{12}$	$-\frac{\sigma(-\sigma+1)}{4}$	$-\frac{\sigma}{2}$	$\frac{1}{2}$
0	$-\frac{\sigma(-\sigma+1)(-\sigma+2)}{12}$	$-\frac{\sigma(-\sigma+1)}{4}$	$-\frac{\sigma}{2}$	$\frac{1}{2}$	0

Similarly, $G_y{}^\sigma$ can be calculated as follows:

$$G_y{}^\sigma = \frac{1}{2}\left(\frac{\partial^\sigma\left(f\left(x-1,\,y+1\right)\right)}{\partial y^\sigma} - \frac{\partial^\sigma\left(f\left(x,\,y\right)\right)}{\partial y^\sigma}\right), \tag{6.12}$$

with its two terms expressed as:

$$\frac{\partial^\sigma\left(f\left(x-1,\,y+1\right)\right)}{\partial y^\sigma} = f\left(x-1,\,y+1\right) - \sigma f\left(x-1,\,y\right)$$
$$-\frac{\sigma\left(-\sigma+1\right)}{2}f\left(x-1,\,y-1\right) \tag{6.13}$$
$$-\frac{\sigma\left(-\sigma+1\right)\left(-\sigma+2\right)}{6}f\left(x-1,\,y-2\right),$$

and

$$\frac{\partial^\sigma\left(f\left(x,\,y\right)\right)}{\partial y^\sigma} = -f\left(x,\,y\right) + \sigma f\left(x,\,y-1\right) + \frac{\sigma\left(-\sigma+1\right)}{2}f\left(x,\,y-2\right)$$
$$+\frac{\sigma\left(-\sigma+1\right)\left(-\sigma+2\right)}{6}f\left(x,\,y-3\right). \tag{6.14}$$

The resulting filter mask I_y that yields $G_y{}^\sigma$ when convolved with the image $f(x, y)$ is thus (Table 6.5):

With Prewitt kernels, gradient components G_x and G_y can be expressed as:

$$G_x = -f\left(x-1,\,y-1\right) - f\left(x-1,\,y\right) - f\left(x-1,\,y+1\right) + f\left(x+1,\,y-1\right)$$
$$+ f\left(x+1,\,y\right) + f\left(x+1,\,y+1\right), \tag{6.15}$$

and

$$G_y = -f(x-1, y-1) - f(x, y-1) - f(x+1, y-1) + f(x-1, y+1)$$
$$+ f(x, y+1) + f(x+1, y+1).$$

(6.16)

Applying G-L formulas in (6.5) and (6.6)–(6.15) leads to $G_x{}^\sigma$ as expressed in (6.17):

$$G_x{}^\sigma = -\frac{1}{2}\frac{\partial^\sigma f(x-1, y-1)}{\partial x^\sigma} - \frac{1}{2}\frac{\partial^\sigma f(x-1, y)}{\partial x^\sigma} - \frac{1}{2}\frac{\partial^\sigma f(x-1, y+1)}{\partial x^\sigma}$$
$$+ \frac{1}{2}\frac{\partial^\sigma f(x+1, y-1)}{\partial x^\sigma} + \frac{1}{2}\frac{\partial^\sigma f(x+1, y)}{\partial x^\sigma} + \frac{1}{2}\frac{\partial^\sigma f(x+1, y+1)}{\partial x^\sigma}.$$

(6.17)

Developing the different terms in (6.17) yields:

$$-\frac{1}{2}\frac{\partial^\sigma f(x-1, y-1)}{\partial x^\sigma} = -\frac{1}{2}f(x-1, y-1) + \frac{\sigma}{2}f(x-2, y-1)$$
$$+ \frac{\sigma(-\sigma+1)}{4}f(x-3, y-1)$$
$$+ \frac{\sigma(-\sigma+1)(-\sigma+2)}{12}f(x-4, y-1),$$

(6.18)

$$-\frac{1}{2}\frac{\partial^\sigma f(x-1, y)}{\partial x^\sigma} = -\frac{1}{2}f(x-1, y) + \frac{\sigma}{2}f(x-2, y) + \frac{\sigma(-\sigma+1)}{4}f(x-3, y)$$
$$+ \frac{\sigma(-\sigma+1)(-\sigma+2)}{12}f(x-4, y),$$

(6.19)

$$-\frac{1}{2}\frac{\partial^\sigma f(x-1, y+1)}{\partial x^\sigma} = -\frac{1}{2}f(x-1, y+1) + \frac{\sigma}{2}f(x-2, y+1)$$
$$+ \frac{\sigma(-\sigma+1)}{4}f(x-3, y+1)$$
$$+ \frac{\sigma(-\sigma+1)(-\sigma+2)}{12}f(x-4, y+1),$$

(6.20)

$$\frac{1}{2}\frac{\partial^\sigma f(x+1,y-1)}{\partial x^\sigma} = \frac{1}{2}f(x+1,y-1) - \frac{\sigma}{2}f(x,y-1)$$
$$-\frac{\sigma(-\sigma+1)}{4}f(x-1,y-1) \qquad (6.21)$$
$$-\frac{\sigma(-\sigma+1)(-\sigma+2)}{12}f(x-2,y-1),$$

$$\frac{1}{2}\frac{\partial^\sigma f(x+1,y)}{\partial x^\sigma} = \frac{1}{2}f(x+1,y) - \frac{\sigma}{2}f(x,y) - \frac{\sigma(-\sigma+1)}{4}f(x-1,y)$$
$$-\frac{\sigma(-\sigma+1)(-\sigma+2)}{12}f(x-2,y),$$
$$(6.22)$$

$$\frac{1}{2}\frac{\partial^\sigma f(x+1,y+1)}{\partial x^\sigma} = \frac{1}{2}f(x+1,y+1) - \frac{\sigma}{2}f(x,y+1)$$
$$-\frac{\sigma(-\sigma+1)}{4}f(x-1,y+1) \qquad (6.23)$$
$$-\frac{\sigma(-\sigma+1)(-\sigma+2)}{12}f(x-2,y+1).$$

As for the representation of G_y^σ, it can be expressed as:

$$G^\sigma{}_y = -\frac{1}{2}\frac{\partial^\sigma f(x-1,y-1)}{\partial y^\sigma} - \frac{1}{2}\frac{\partial^\sigma f(x,y-1)}{\partial y^\sigma} - \frac{1}{2}\frac{\partial^\sigma f(x+1,y-1)}{\partial y^\sigma}$$
$$+\frac{1}{2}\frac{\partial^\sigma f(x-1,y+1)}{\partial x^\sigma} + \frac{1}{2}\frac{\partial^\sigma f(x,y+1)}{\partial y^\sigma} + \frac{1}{2}\frac{\partial^\sigma f(x+1,y+1)}{\partial y^\sigma},$$
$$(6.24)$$

with its different terms expressed as follows:

$$-\frac{1}{2}\frac{\partial^\sigma f(x-1,y-1)}{\partial y^\sigma} = -\frac{1}{2}f(x-1,y-1) + \frac{\sigma}{2}f(x-1,y-2)$$
$$+\frac{\sigma(-\sigma+1)}{4}f(x-1,y-3) \qquad (6.25)$$
$$+\frac{\sigma(-\sigma+1)(-\sigma+2)}{12}f(x-1,y-4),$$

$$-\frac{1}{2}\frac{\partial^\sigma f(x, y-1)}{\partial y^\sigma} = -\frac{1}{2}f(x, y-1) + \frac{\sigma}{2}f(x, y-2)$$

$$+ \frac{\sigma(-\sigma+1)}{4}f(x, y-3) \qquad (6.26)$$

$$+ \frac{\sigma(-\sigma+1)(-\sigma+2)}{12}f(x, y-4),$$

$$-\frac{1}{2}\frac{\partial^\sigma f(x+1, y-1)}{\partial y^\sigma} = -\frac{1}{2}f(x+1, y-1) + \frac{\sigma}{2}f(x+1, y-2)$$

$$+ \frac{\sigma(-\sigma+1)}{4}f(x+1, y-3) \qquad (6.27)$$

$$+ \frac{\sigma(-\sigma+1)(-\sigma+2)}{12}f(x+1, y-4),$$

$$\frac{1}{2}\frac{\partial^\sigma f(x-1, y+1)}{\partial y^\sigma} = \frac{1}{2}f(x-1, y+1) - \frac{\sigma}{2}f(x-1, y)$$

$$- \frac{\sigma(-\sigma+1)}{4}f(x-1, y-1) \qquad (6.28)$$

$$- \frac{\sigma(-\sigma+1)(-\sigma+2)}{12}f(x-1, y-2),$$

$$\frac{1}{2}\frac{\partial^\sigma f(x, y+1)}{\partial y^\sigma} = \frac{1}{2}f(x, y+1) - \frac{\sigma}{2}f(x, y) - \frac{\sigma(-\sigma+1)}{4}f(x, y-1)$$

$$- \frac{\sigma(-\sigma+1)(-\sigma+2)}{12}f(x, y-2),$$

$$(6.29)$$

$$\frac{1}{2}\frac{\partial^\sigma f(x+1, y+1)}{\partial y^\sigma} = \frac{1}{2}f(x+1, y+1) - \frac{\sigma}{2}f(x+1, y)$$

$$- \frac{\sigma(-\sigma+1)}{4}f(x+1, y-1) \qquad (6.30)$$

$$- \frac{\sigma(-\sigma+1)(-\sigma+2)}{12}f(x+1, y-2).$$

Therefore, filter masks I_x and I_y used to compute G_x^σ and G_y^σ based on initial Prewitt kernels can be expressed as shown in Tables 6.6 and 6.7, respectively.

Table 6.6 Fractional order Prewitt mask I_x

Offset : $x \downarrow y \rightarrow$	-1	0	1
-4	$\frac{\sigma(-\sigma+1)(-\sigma+2)}{12}$	$\frac{\sigma(-\sigma+1)(-\sigma+2)}{12}$	$\frac{\sigma(-\sigma+1)(-\sigma+2)}{12}$
-3	$\frac{\sigma(-\sigma+1)}{4}$	$\frac{\sigma(-\sigma+1)}{4}$	$\frac{\sigma(-\sigma+1)}{4}$
-2	$\frac{\sigma}{2} - \frac{\sigma(-\sigma+1)(-\sigma+2)}{12}$	$\frac{\sigma}{2} - \frac{\sigma(-\sigma+1)(-\sigma+2)}{12}$	$\frac{\sigma}{2} - \frac{\sigma(-\sigma+1)(-\sigma+2)}{12}$
-1	$-\frac{1}{2} - \frac{\sigma(-\sigma+1)}{4}$	$-\frac{1}{2} - \frac{\sigma(-\sigma+1)}{4}$	$-\frac{1}{2} - \frac{\sigma(-\sigma+1)}{4}$
0	$-\frac{\sigma}{2}$	$-\frac{\sigma}{2}$	$-\frac{\sigma}{2}$
1	$\frac{1}{2}$	$\frac{1}{2}$	$\frac{1}{2}$

Table 6.7 Fractional order Prewitt mask I_y

Offset : $x \downarrow y \rightarrow$	-4	-3	-2	-1	0	1
-1	$\frac{\sigma(-\sigma+1)(-\sigma+2)}{12}$	$\frac{\sigma(-\sigma+1)}{4}$	$\frac{\sigma}{2} - \frac{\sigma(-\sigma+1)(-\sigma+2)}{12}$	$-\frac{1}{2} - \frac{\sigma(-\sigma+1)}{4}$	$-\frac{\sigma}{2}$	$\frac{1}{2}$
0	$\frac{\sigma(-\sigma+1)(-\sigma+2)}{12}$	$\frac{\sigma(-\sigma+1)}{4}$	$\frac{\sigma}{2} - \frac{\sigma(-\sigma+1)(-\sigma+2)}{12}$	$-\frac{1}{2} - \frac{\sigma(-\sigma+1)}{4}$	$-\frac{\sigma}{2}$	$\frac{1}{2}$
1	$\frac{\sigma(-\sigma+1)(-\sigma+2)}{12}$	$\frac{\sigma(-\sigma+1)}{4}$	$\frac{\sigma}{2} - \frac{\sigma(-\sigma+1)(-\sigma+2)}{12}$	$-\frac{1}{2} - \frac{\sigma(-\sigma+1)}{4}$	$-\frac{\sigma}{2}$	$\frac{1}{2}$

Table 6.8 Fractional order Sobel mask I_x

Offset : $x \downarrow y \rightarrow$	-1	0	1
-4	$\frac{\sigma(-\sigma+1)(-\sigma+2)}{12}$	$\frac{\sigma(-\sigma+1)(-\sigma+2)}{12}$	$\frac{\sigma(-\sigma+1)(-\sigma+2)}{12}$
-3	$\frac{\sigma(-\sigma+1)}{4}$	$\frac{\sigma(-\sigma+1)}{2}$	$\frac{\sigma(-\sigma+1)}{4}$
-2	$\frac{\sigma}{2} - \frac{\sigma(-\sigma+1)(-\sigma+2)}{12}$	$\sigma - \frac{\sigma(-\sigma+1)(-\sigma+2)}{6}$	$\frac{\sigma}{2} - \frac{\sigma(-\sigma+1)(-\sigma+2)}{12}$
-1	$-\frac{1}{2} - \frac{\sigma(-\sigma+1)}{4}$	$-1 - \frac{\sigma(-\sigma+1)}{2}$	$-\frac{1}{2} - \frac{\sigma(-\sigma+1)}{4}$
0	$-\frac{\sigma}{2}$	$-\sigma$	$-\frac{\sigma}{2}$
1	$\frac{1}{2}$	1	$\frac{1}{2}$

As for Sobel method, the calculations are similar to Prewitt method computations. Thus, we will present the forms of the masks I_x and I_y respectively in Tables 6.8 and 6.9 for the components G_x (6.31) and G_y (6.32):

$$G_x = -f(x-1, y-1) - 2f(x-1, y) - f(x-1, y+1) + f(x+1, y-1) \\ + 2f(x+1, y) + f(x+1, y+1),$$

$$(6.31)$$

$$G_y = -f(x-1, y-1) - 2f(x, y-1) - f(x+1, y-1) + f(x-1, y+1) \\ + 2f(x, y+1) + f(x+1, y+1).$$

$$(6.32)$$

Table 6.9 Fractional order Sobel mask I_y

$Offset : x \downarrow y \rightarrow$	-4	-3	-2		-1		0	1
-1	$\frac{\sigma(-\sigma+1)(-\sigma+2)}{12}$	$\frac{\sigma(-\sigma+1)}{4}$	$\frac{\sigma}{2} -$	$\frac{\sigma(-\sigma+1)(-\sigma+2)}{12}$	$-\frac{1}{2} -$	$\frac{\sigma(-\sigma+1)}{4}$	$-\frac{\sigma}{2}$	$\frac{1}{2}$
0	$\frac{\sigma(-\sigma+1)(-\sigma+2)}{6}$	$\frac{\sigma(-\sigma+1)}{2}$	$\sigma -$	$\frac{\sigma(-\sigma+1)(-\sigma+2)}{6}$	$-1 -$	$\frac{\sigma(-\sigma+1)}{2}$	$-\sigma$	1
1	$\frac{\sigma(-\sigma+1)(-\sigma+2)}{12}$	$\frac{\sigma(-\sigma+1)}{4}$	$\frac{\sigma}{2} -$	$\frac{\sigma(-\sigma+1)(-\sigma+2)}{12}$	$-\frac{1}{2} -$	$\frac{\sigma(-\sigma+1)}{4}$	$-\frac{\sigma}{2}$	$\frac{1}{2}$

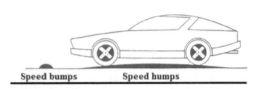

Fig. 6.2 Speeds humps, bumps, and cushions

Road Obstacle Detection

As this chapter deals with the identification of road obstacles, the edge detection technique has been chosen in order to realize this task. The different road obstacles will be first classified then a comparative study between the integer order and the fractional-order Sobel edge detection method will be presented.

Speed humps, bumps, and cushions are vertical elevations of the road level made for the vehicle motion to disturb the driver and force him to pull the brakes in order to slow down the speed of his vehicle [71].

Speed humps have a length larger than the distance between the front wheels and the back wheels. In addition to that, their elevation increases and decreases gradually. The top of this traffic calming can be either flat or round.

Speed bumps are more destructive, and they present a length that is much smaller than the wheelbase. This type of traffic calming is not recommended due to its effect on the vehicles which may cause some damages at the mechanical level of the vehicle.

As for cushions, they are designed in a way that the emergency cars can pass over without slowing their speed, they target only the passenger's vehicles. In other words, cushions and humps are similar except that cushions take into consideration the emergency vehicles. Figure 6.2 shows the three speed delimiters.

Three images shown in Fig. 6.3 are taken into consideration representing the types mentioned before.

Each one of them will pass through four different processing steps presented in the flowchart in Fig. 6.4 and the result of each step will be presented separately.

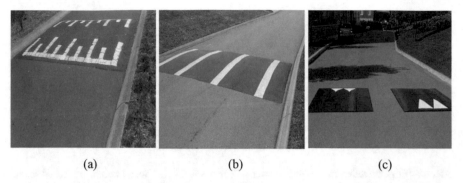

(a)   (b)   (c)

Fig. 6.3 Three different types of traffic control devices. (**a**) Speed hump, (**b**) Speed bump, and (**c**) Speed cushion

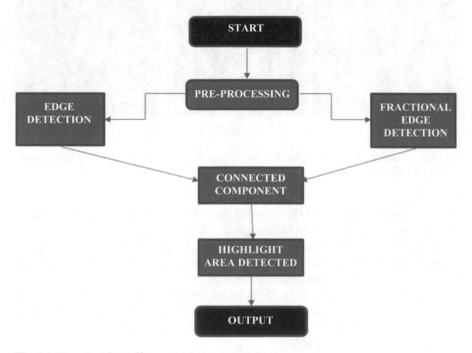

Fig. 6.4 Flowchart for traffic control device detection

The first step is a preprocessing work executed on the original image. The initial RGB image is converted into a gray scale. In addition to that, we apply a noise reduction algorithm on the image using an averaging filter in order to obtain better results in the upcoming processing steps.

As for the second step, it is a direct edge detection applied on the resulting image. The already presented masks for Sobel's method will be applied. As for

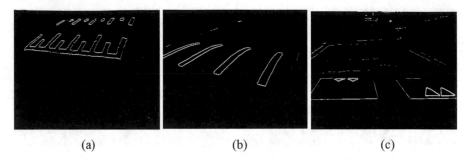

Fig. 6.5 Edge detection using integer order of Sobel's mask. (**a**) Speed hump, (**b**) Speed bump, and (**c**) Speed cushion

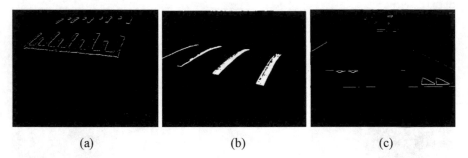

Fig. 6.6 Edge detection using fractional order of Sobel's mask. (**a**) Speed hump, (**b**) Speed bump, and (**c**) Speed cushion

the fractional mask, the value for σ was chosen after several trials, and it was set to be equal to 0.85 (for humps and cushions) or 0.6 (for bumps).

Concerning the integer edge detection method, the results of the three obstacles, presented in Fig. 6.3, are shown in the Fig. 6.5. As for the fractional edge detection technique, the results are presented in Fig. 6.6.

Now for the third step, edge pixels which are not part of the area of interest and false edges will be removed. To do that, connected components are detected, while very small components as well as non-connected components are ignored (i.e., marked as non-edge pixels).

Figure 6.7 shows the results of the integer method after detecting connected components and ignoring small objects, whereas Fig. 6.8 shows the outcome of the fractional method for the three images.

The final step is to highlight the remaining white pixels in order to define the area of interest. This domain is highlighted for the six figures of both algorithms by using red box surrounding them.

The final result for each type in the fractional and integer methods is represented in Figs. 6.9, 6.10, and 6.11 that show the highlighted area for the speed hump, bump, and cushion respectively.

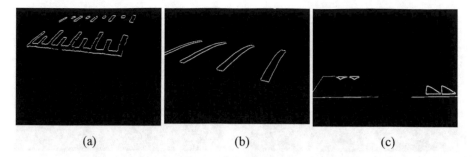

Fig. 6.7 Connected component result for the integer method. (**a**) Speed hump, (**b**) Speed bump, and (**c**) Speed cushion

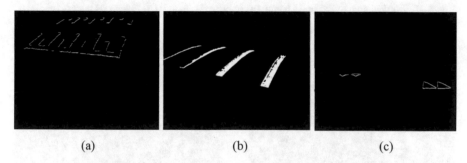

Fig. 6.8 Connected component result for the fractional method. (**a**) Speed hump, (**b**) Speed bump, and (**c**) Speed cushion

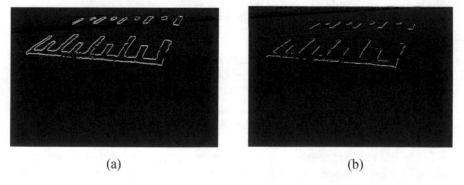

Fig. 6.9 Marked area for the speed hump. (**a**) Integer result, and (**b**) Fractional result

Results Discussion

Comparing the integer and fractional Sobel methods when detecting edges, one can observe that the fractional method results in significantly less false edges.

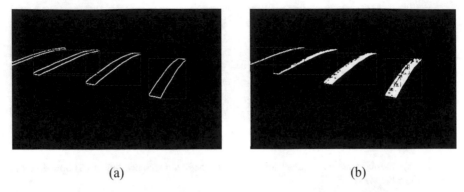

(a)   (b)

Fig. 6.10 Marked area of the speed bump. (**a**) Integer result, and (**b**) Fractional result

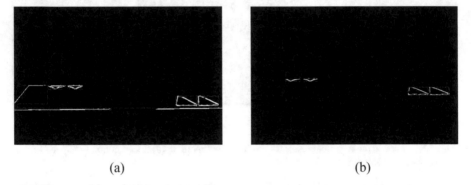

(a)   (b)

Fig. 6.11 Marked area of the speed cushion. (**a**) Integer result, and (**b**) Fractional result

The result in Fig. 6.5a is more accurate in detecting the shape on the hump than the fractional method shown in the Fig. 6.6a, but, in both cases, the painting on the speed hump can be clearly observed.

When it comes to the speed bump, one can see that the fractional detection shown in Fig. 6.6b presents very thick edges, an effect similar to image binarization while keeping only the regions of interest.

In the last case representing the speed cushion, one can see that in both cases shown in Figs. 6.5c and 6.6c, the triangle shape on the speed cushion indicating road direction is successfully highlighted.

After evaluating the results of these two edge detection approaches, their complexity is estimated in terms of the processing time required for the simulator to output an edge image, once an input image is available. This estimation is performed since edge detection might be applied in a real time scenario, where edges are detected as soon as the image is acquired in order to be used in subsequent operations, such as controlling the braking system of a vehicle in motion. The processing time for the speed bump, hump, and cushion according to both methods are represented in Table 6.10.

Table 6.10 Processing time for bump, hump, and cushion detection

	Integer edge detection (in seconds)	Fractional edge detection (in seconds)
Speed bump	3.613338	3.600586
Speed hump	3.868696	3.775127
Speed cushion	3.643436	3.558770

Looking at Table 6.10, it is clear that fractional edge detection gives a faster result comparing to integer edge detection for all the three types of traffic calming devices used in this study. According to these results, fractional edge detection presents an average of 3.644828 s, while the integer method has an average of 3.70849 s. According to the speed average of these two methods, the fractional edge detection is faster than the integer method by 1.72%. It is important to mention that the numbers shown in Table 6.10 depend on the size of the images, as well as on the hardware platform and software tools used to perform the computations. Therefore, considering each value by itself does not have a meaningful interpretation; it is rather the relative time taken for one method to be completed compared to the other that counts, given the same images and computational platforms.

Applications

Edge detection techniques, in particular of fractional order, could be used for wider applications where the detection of specific points and areas of the images are required. In addition to that, the applications relying on faster analysis and calculations can refer to the fractional order methods.

On the other hand, the detection of road obstacles can serve for further applications, mainly on cars or any automotive vehicle. Thus, this module can be connected to the braking system such that, once such abnormalities are detected, a distance calculation is proceeded in order to act on the brakes such that the car passes through them smoothly.

Another application can be the speed control of the car. Once an obstacle is detected, the speed controller may alter the car speed and acceleration in order to reduce fuel consumption.

Thus, the obstacle detection system can be implemented in several applications, mainly in autonomous car and self-driving vehicles, in order to act on the braking, the fuel consumption, the chassis behavior, and so on. Some existent works and videos, published by big car manufacturers as Ford, Continental, and others, have already shown the importance of such embedded systems [72].

Conclusion and Future Work

In this chapter, an edge detection method was designed based on fractional calculus, and applied in the context of road obstacle detection for smarter transportation and safer cities.

Concerning the results, the fractional detection method has shown better performance in terms of computational load, with a gain of 1.25–2.33% in simulation time compared to the integer-based method. Furthermore, the fractional order method yielded thicker edges, a desired property that allows for better detection of obstacles.

As for the applications, road obstacle detection can be applied to the car controlling unit in order to regulate its speed and to optimize its braking system, concerning the timing and severity of the braking. It can be also implemented within autonomous cars to control the suspension as well as the cruise control unit.

Thus, the automatic identification of these obstacles would gain a great deal of importance if exploited in the context of autonomous driverless vehicles and safe cities.

As for the future works, the automatic calculation of the best value of σ would be of great importance in order, not only to enhance the image characteristics but also to reduce the processing time. In addition, the implementation of the fractional calculus on Laplacian methods is also very important. In the end, the live streaming and video processing would be very important in order to adjust the car's speed in real time.

References

1. Yaacoub, C., & Sarkis, M. (2017). Systematic polar codes for joint source-channel coding in wireless sensor networks and the internet of things. *Procedia Computer Science, 110*, 266–273.
2. Al-Turjman, F. (2019). *Intelligence in IoT-enabled smart cities.* s.l.: CRC Press.
3. Al-Turjman, F. (2018). QoS–aware data delivery framework for safety-inspired multimedia in integrated vehicular-IoT. *Computer Communications Journal, 121*, 33–43.
4. Al-Turjman, F., & Alturjman, S. (2018). Confidential smart-sensing framework in the IoT era. *Journal of Supercomputing, 74*(10), 5187–5198.
5. Chan, C.-Y. (2017). Advancements, prospects, and impacts of automated driving systems. *International Journal of Transportation Science and Technology, 6*(3), 208–216.
6. Canat, S., & Faucher, J. (2005). Modeling, identification and simulation of induction machine with fractional derivative. In U-Books (Ed.), *Fractional differentiation and its applications* (pp. 459–470). s.l.: s.n.
7. Coman, S., Comnac, V., Boldisor, C., & Dumitrache, D. (2010). *Fractional order control for DC electrical drives in networked control systems.* Brasov, Romania: s.n.
8. Agrawal, O. (2004). Application of fractional derivatives in thermal analysis of disk brakes. *Journal of Nonlinear Dynamics, 38*, 191–206.
9. Moreau, X., Ramus-Serment, C., & Oustaloup, A. (2002). Fractional differentiation in passive vibration control. *Journal of Nonlinear Dynamics, 29*, 343–362.
10. Benchellal, A., Poinot, T., & Trigeassou, J.-C. (2006). Approximation and identification of diffusive interfaces by fractional models. *Signal Processing, 86*(10), 2712–2727.

11. Abi Zeid Daou, R., Moreau, X., Assaf, R., & Christohpy, F. (2012). *Analysis of HTE fractional order system in the thermal diffusive interface - Part 1: Application to a semi-infinite plane medium*. Lebanon: s.n.
12. Benchellal, A., Poinot, T., & Trigeassou, J.-C. (2008). Fractional modelling and identification of a thermal process. *Journal of Vibration and Control, 14*(9/10), 1403–1414.
13. Adhikari, P., Karmakar, A., & Das, R. (2015). *A switched capacitor based realization of fractional order low-pass filters*. Gwalior, India: s.n.
14. Gonzalez, E., & Petras, I. (2015). *Advances in fractional calculus: Control and signal processing applications*. Szilvasvarad, Hungary: s.n.
15. Ortigueira, M., Machado, J.-A., Trujillo, J., & Vinagre, B. (2011). Advances in fractional signals and systems. *Signal Processing, 91*(3), 349.
16. Abi Zeid Daou, R., & Moreau, X. (2015). *Fractional Calculus: Applications*. New York: Nova.
17. Miller, K., & Ross, B. (1993). *An introduction to the fractional calculus and fractional differential equations*. New York: Wiley.
18. Oldham, K., & Spanier, J. (1974). *The fractional calculus*. New York: Academic Press.
19. Oustaloup, A. (1975). *Etude et Réalisation d'un systme d'asservissement d'ordre 3/2 de la fréquence d'un laser à colorant continu*. Bordeaux, France: Universitu of Bordeaux.
20. Amoako-Yirenkyi, P., Appati, J., & Dontwi, I. (2016). A new construction of a fractional derivative mask for image edge analysis based on Riemann-Liouville fractional derivative. *Advances in Difference Equations, 1*(1), 1–23.
21. Hacini, M., Hacini, A., Akdag, H., & Hachouf, F. (2017). *A 2D-fractional derivative mask for image feature edge detection*. Fez, Morocco: s.n.
22. Kamaruddin, N., Abdullah, N., & Ibrahim, R. (2015). Image segmentation based on fractional non-Markov poisson stochastic process. *Pakistan Journal of Statistics, 31*(5), 557–574.
23. Wang, J., Song, Q., Jiang, Z., & Zhou, Z. (2016a). *A novel InSAR based off-road positive and negative obstacle detection technique for unmanned ground vehicle*. Beijing, China: s.n.
24. Wang, Z., Su, J., & Zhang, P. (2016b). *Image edge detection algorithm based onwavelet fractional differential theory*. Chengdu, China: s.n.
25. Xu, Z., Zhuang, Y., & Chen, H. (2006). *Obstacle detection and road following using laser scanner*. Dalian, China: s.n.
26. Yalcin, O., Sayar, A., Arar, O. F., Apinar, S., & Kosunalp, S. (2014). *Detection of road boundaries and obstacles using LIDAR*. Colchester, UK: s.n.
27. Abi Zeid Daou, R., & Moreau, X. (2014). *Fractional calculus: Theory*. New York: Nova Science Publishers Inc.
28. Assaf, R., Moreau, X., Abi Zeid Daou, R., & Christohpy, F. (2012). *Analysis of Hte fractional order system in hte thermal diffusive interface - Part 2: application to a finite medium*. Lebanon: s.n.
29. Trigeassou, J.-C., Poinot, T., Lin, J., Oustaloup, A., & Levron, F. (1999). *Modeling and identification of a non integer order system*. Karlsruhe, Germany: IFAC.
30. Jith, J., & Sarkar, S. (2018). Boundary layer impedance model to analyse the visco-thermal acousto-elastic interactions in centrifugal compressors. *Journal of Fluids and Structures, 81*, 179–200.
31. Tavazoei, M. S., & Haeri, M. (2008). Regular oscillations or chaos in a fractional order system with any effective dimension. *Nonlinear Dynamics, 54*(3), 213–222.
32. Daftardar-Gejji, V., & Bhalekar, S. (2010). Chaors in fractional ordered Liu system. *Computers & Mathematics with Applications, 59*, 1117–1127.
33. Ionescu, C., Machado, J., & de Keyser, R. (2011). Modeling of the lung impedance using a fractional-order ladder network with constant phase elements. *IEEE Transactions on Biomedical Circuits and Systems, 5*(1), 83–89.
34. Melchior, P., Pellet, M., Petit, J., Cabelguen, J. M., & Oustaloup, A. (2012). Analysis of muscle length effect on an S type motor-unit fractional multi-model. *Signal, Image and Video Processing, 6*(3), 421–428.
35. Hennion, M., & Hanert, E. (2013). How to avoid unbounded drug accumulation with fractional pharmacokinetics. *Journal of Pharmacokinetics and Pharmacodynamics, 40*, 691–700.

36. Oustaloup, A. (1991). *La commande CRONE*. Paris: Hermes.
37. Charef, A., & Fergani, N. (2010). *$PI\lambda D\mu$ controller tuning for desired closed-loop response using impulse response*. Spain: s.n.
38. Zhang, B., Pi, Y., & Luo, Y. (2012). Fractional order sliding-mode control based on parameters auto-tuning for velocity control of permanent magnet synchronous motor. *ISA Transactions, 51*(5), 649–656.
39. Ladacia, S., Loiseaua, J., & Charefb, A. (2008). Fractional order adaptive high-gain controllers for a class of linear systems. *Communication in Nonlinear Science and Numerical Simulation, 13*(4), 707–714.
40. Chen, Y., Vinagre, B., & Podlubny, I. (2004). Fractional order disturbance observer for robust vibration suppression. *Nonlinear Dynamuics, 38*, 355–367.
41. Oustaloup, A. (1995). *La dérivation non entière: Théorie, synthèse et applications*. Paris: Hermes.
42. Abi Zeid Daou, R., Francis, C., & Moreau, X. (2009). Synthesis and implementation of non-integer integrators using RLC devices. *International Journal of Electronics, 96*(12), 1207–1223.
43. Ramus-Serment, C., Moreau, X., Nouillant, M., Oustaloup, A., & Levron, F. (2002). Generalised approach on fractional response of fractal networks. *Journal of Chaos, Solitons and Fractals, 14*, 479–488.
44. Veeraian, P., Gandhi, U., & Mangalanathan, U. (2018). Design and analysis of fractional order seismic transducer for displacement and acceleration measurements. *Journal of Sound and Vibration, 419*, 123–139.
45. Germoso, C., Fraile, A., Alarcon, E., Aguado, J. V., & Chinesta, F. (2017). From standard to fractional structural visco-elastodynamics: Application to seismic site response. *Physics and Chemistry of the Earth, 98*, 3–15.
46. Blackledge, J. (2008). Application of the fractal market hypothesis for modelling macroeconomic time series. *ISAST Transactions on Electronics and Signal Processing, 1*(2), 1–22.
47. Cartea, A., & del-Castillo-Negrete, D. (2007). Fractional diffusion models of option prices in markets with jumps. *Physica A, 374*, 749–763.
48. Tenreiro Machado, J., Kiryakova, V., & Mainardi, F. (2011). Recent history of fractional calculus. *Commun Nonlinear Sci Numer Simulat, 16*, 1140–1153.
49. Hennelly, B., & Sheridana, J. (2003). Image encryption and the fractional Fourier transform. *Optik, 114*(6), 251–265.
50. Şamil Yetik, I., Alper Kutay, M., & Ozaktasc, H. (2001). Image representation and compression with the fractional Fourier transform. *Optics Communications, 197*(4–6), 275–278.
51. Janev, M., et al. (2011). Fully fractional anisotropic diffusion for image denoising. *Mathematical and Computer Modelling, 54*(1–2), 729–741.
52. Cuesta, E., Kirane, M., & Malik, S. (2012). Image structure preserving denoising using generalized fractional time integrals. *Signal Processing, 92*(2), 553–563.
53. Ghamisi, P., Couceiro, M., Benediktsson, J., & Ferreira, N. (2012). An efficient method for segmentation of images based on fractional calculus and natural selection. *Expert Systems with Applications, 39*(16), 12407–12417.
54. Mathieu, B., Melchior, P., Oustaloup, A., & Ceyral, C. (2003). Fractional differentiation for edge detection. *Signal Processing, 83*(11), 2421–2432.
55. Yang, H., Ye, Y., Wang, D., & Jiang, B. (2010). *A novel fractional-order signal processing based edge detection method*. Singapore: s.n.
56. ElAraby, W., Madian, A. H., Ashour, M. A., Farag, I., & Nassef, M. (2017). *Fractional edge detection based on genetic algorithm*. Beirut, Lebanon: s.n.
57. ElAraby, W., Median, A. H., Ashour, M. A., Farag, I., & Nassef, M. (2016). *Fractional canny edge detection for biomedical applications*. Giza, Egypt: s.n.
58. Tian, D., Wu, J., & Yang, Y. (2014). *A fractional-order edge detection operator for medical image structure feature extraction*. Changsha, China: s.n.
59. Ismail, S., Radwan, A., Madian, A., & Abu-ElYazeed, M. (2016). *Comparative study of fractional filters for Alzheimer disease detection on MRI images*. Vienna, Austria: s.n.

60. Discant, A., Emerich, S., Lupu, E., Rogozan, A., & Bensrhair, A. (2007). *Ruttier obstacle classification by use of fractional B-spline wavelets and moments*. Warsaw, Poland: s.n.
61. Chen, Y., Wang, Z., & Moore, K. (2006). *Optimal spraying control of a diffusion process using mobile actuator networks with fractional potential field based dynamic obstacle avoidance*. Ft. Lauderdale, FL: s.n.
62. Folorunso, O., & Vincent, R. (2009). *A descriptive algorithm for sobel image edge detection*. Macon, GA: s.n.
63. Muthukrishnan, R., & Radha, M. (2011). Edge detection techniques for image segmentation. *International Journal of Computer Science and Information Technology (IJCSIT), 3*(6), 259.
64. Maini, R., & Aggarwal, H. (2009). Study and comparison of various image edge detection techniques. *International Journal of Image Processing (IJIP), 3*(1), 1–11.
65. Shrivakshan, G., & Chandrasekar, C. (2012). A comparison of various edge detection techniques used in image processing. *International Journal of Computer Science Issues, 9*(5), 269–276.
66. Popa, B. (2017). *Study about the edge detection algorithm and its applications*. Sinaia, Romania: s.n.
67. Katiyar, S., & Arun, P. (2012). Comparative analysis of common edge detection techniques in context of object extraction. *IEEE Transactions on Geoscience and Remote Sensing (TGRS), 20*(11), 68–78.
68. Yang, L., Wu, X., Zhao, D., Li, H., & Zhai, J. (2011). *An improved Prewitt algorithm for edge detection based on noised image*. Shanghai, China: s.n.
69. Karla, A., & Lal Chhokar, R. (2016). International conference on micro-electronics and telecommunication engineering (ICMETE). In *A hybrid approach using Sobel and canny operator for digital image edge detection*. Ghaziabad, India: s.n.
70. Selvakumar, P., & Hariganesh, S. (2016). *The performance analysis of edge detection algorithms for image processing*. Kovilpatti, India: s.n.
71. Berthod, C. (2011). *Traffic calming speed humps and speed cushions*. Edmonton, AL: Ministère des Transports du Québec.
72. Murray, M. (2016). *Potholes no match for new ford model*. Detroit, MI: Elephant Insurance.

Chapter 7
Machine Learning Parameter Estimation in a Smart-City Paradigm for the Medical Field

M. Bhuvaneswari, G. Naveen Balaji, and F. Al-Turjman

Introduction

A smart city uses the principles of intelligent technological infrastructure in city management for different purposes, such as planning, analysis, and improving the quality of services. In this sense, smart cities can significantly benefit from machine learning (ML) and artificial intelligence (AI) techniques when providing smart management, smart transportation, smart economy, and smart health care applications [1–4]. Especially in the health care sector, there have been several attempts toward further automation and ML applications. For example, Rose et al. [5] used convolutional neural networks (CNNs) to create a new network architecture with the aim of multi-channel data acquisition and supervised feature learning. Extracting features from brain images (e.g., magnetic resonance imaging [MRI], functional magnetic resonance imaging [fMRI]) can help with the early diagnosis and prognosis of severe diseases, such as glioma. Moreover, Kuang and He [6] used a deep belief network (DBN) for the classification of mammography images in order to detect calcifications, which might be an indicator of early breast cancer.

ML is a popular and an advanced technology that is now used in all areas of research. Generally, machine learning is classified into two methods of learning: supervised and unsupervised learning methods. As the name suggests, supervised learning methods are those in which the machines have to be trained prior and have to be supervised while performing operations [7]. In unsupervised learning, the system or machine can work under real-time problems or scenarios; they do not

M. Bhuvaneswari · G. N. Balaji (✉)
Department of ECE, SNS College of Technology, Coimbatore, Tamil Nadu, India
e-mail: yoursgnb@gmail.com

F. Al-Turjman
Department of Computer Engineering, Antalya Bilim University, Antalya, Turkey

© Springer Nature Switzerland AG 2020
F. Al-Turjman (ed.), *Smart Cities Performability, Cognition, & Security*,
EAI/Springer Innovations in Communication and Computing,
https://doi.org/10.1007/978-3-030-14718-1_7

have to be trained about the operation that is going to be performed. Supervised learning methods [8] can be further classified into either classification techniques or regression techniques. Classification techniques are mostly used when a set of data has to be classified according to the required parameters [9]. Regression techniques are mostly used when the output must be a real value.

In the medical field, images are captured through various technologies, such as computed tomography (CT), MRI, and mammography. With these methods, locating an abnormal component in a captured image can be laborious. However, improvements in medical image analysis techniques have made it possible to detect even a minute abnormalities, making diagnosis much easier. Gaussian mixture modelling (GMM) performs classifications by extracting global statistics from Gaussian distributions of pixel intensity in an image data set [10]. GMM is especially well suited for parameter estimation and classification because of its implementation facility and efficiency in the representation of data.

With the help of a GMM-based algorithm, features can be extracted using the feature extraction technique [11] and a histogram of the image can be obtained. The histogram is able to provide the density or intensity of the abnormalities that are found in the image because a GMM-based classification is normally carried out with pixel density. A Gaussian distribution is commonly used for its high level of realistic applicability and similar data behaviors [12–15]. In statistical approaches, it is assumed that pixel values follow a particular distribution; hence, a mixture model approach is applied to those values. A common practice is to assume that intensity values follow a Gaussian distribution with two parameters: mean and variance. Well-known approaches for estimating the parameters of a model include maximum likelihood estimation (MLE) via an expectation maximization algorithm (EM) or maximum a posteriori estimation (MAP).

MAP is a likelihood estimation technique that allows for image restoration and classification. It has outstanding performance when comes to parameter estimation because it provides the maximum likelihood function by comparing the range and covering all nearby regions of interest for the convergence of the region of interest [16–20].

With MLE, when data are needed for a particular group, they can be collected from the average of the whole group using mean or variance, among others, as parameters. MLE is used to estimate the parameters for image classification through the EM algorithm [21]. The estimated parameters can be used for the classification of images. When the estimated parameters are used for two different types of classifiers, such as GMM classifiers and support vector machine (SVM) classifiers, they show differences in the accuracy of image classification. The mean, variance, and mixture weights can be used to train both the classifiers.

SVM is a learning vector [22] known for its various applications in almost all areas of research, such as image segmentation or classification and speech processing [23, 24]. SVM-based classifiers are used for their outstanding performance in training machines. The results obtained are comparatively better than other techniques used for the training and classification of images [25]. SVM is known for its kernel classifier, which works well for image classification [26] compared with other normal classifiers based on the parameter estimation.

The proposed work in this chapter includes image classifications carried out by several classifiers and a comparison of the results obtained by the different classifiers. The results indicate that SVM classifiers have a superior level of accuracy. The classifiers were trained with a set of normal and abnormal images. While performing the classification, the different values for the cost functions were recorded. The results show that the classifiers did not perform better for higher function values compared with SVM.

Methodology

The methodology of this work is illustrated in Fig. 7.1, which shows the inputs and methods used for the estimation of the parameters, as well as how these parameters would be used in the training and testing of both classifiers for the classification of medical images. A set of medical images were used for both training and testing, of which 25% of the images were abnormal; the classifier should able to correctly classify the images. The images were taken from different aspects, both normal and abnormal were combined, and they were trained. Then, the rest of the images were used for the testing phase for both classifiers, which were able to correctly classify the abnormal and normal images.

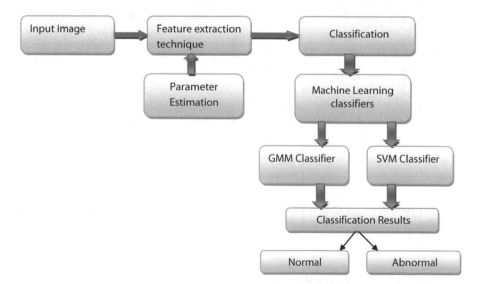

Fig. 7.1 Flow chart of the proposed methodology

Gaussian Mixture Model

GMM is a parametric probability density function represented as a weighted sum of Gaussian component densities. GMMs are commonly used as a parametric model of the probability distribution of continuous measurements [27]. GMM parameters are estimated from training data using the iterative EM algorithm.

A GMM is a weighted sum of M component Gaussian densities as given by the following equation:

$$p(X|\lambda) = \sum_{i=1}^{M} w_i \ g\left(X|\mu_i, \sum_i\right) \tag{7.1}$$

Here, x is a D-dimensional continuous valued data vector (measurement of features), W_i, $i = 1\ldots,M$, are the mixture weights, and $g(x|\mu_i, i)$, $i = 1\ldots,M$, are the component Gaussian densities. The D-variate Gaussian function of the form in each component density is given by the following:

$$g\left(x|\mu i, \sum i\right) = \frac{1}{(2\Pi)^{\left(D/2\right)} \left|\sum_i\right|^{1/2}} \exp\left[-\frac{1}{2}(x-\mu_i)'\sum_i^{-1}(x-\mu_i)\right]$$

$$\tag{7.2}$$

Here, the mean vector is μ_i and the covariance matrix is $\sum_i$.
The constraint of the mixture density is met and given by the following form:

$$\sum_{i=1}^{M} w_i = 1 \tag{7.3}$$

The mean vectors, covariance matrices, and mixture weights from all component densities were used for the complete Gaussian distribution parameterization. These parameters are collectively represented by the following notation:

$$\lambda = \left\{w_i, \mu_i, \sum_i\right\} \quad i = 1\ldots M \tag{7.4}$$

With this, the EM and MLE are used for the estimation of the parameters, as described in the following section.

Maximum Likelihood Parameter Estimation

Given training vectors and a GMM configuration, the objective is to estimate the parameters of the GMM, λ, which in some sense best match the distribution of the training feature vectors. Several techniques are available for estimating the parameters of a GMM, and they have their own form of estimations. Among the various techniques available, the most popular and well-established method is the MLE.

To maximize the likelihood of the GMM, ML estimation of the model parameters is used, which will be better in the training data. For a sequence of T training vectors $X = \{x_1 \ldots ,x_T\}$, the GMM likelihood, assuming independence between the vectors (7.1), can be written as follows:

$$p(x|\lambda) = \prod_{t=1}^{T} P(X_t|\lambda) \tag{7.5}$$

On the EM iteration, estimation formulas are used to guarantee a monotonic increase in the model's likelihood value, as follows:

Mixture weights:

$$\overline{W_i} = \frac{1}{T} \sum_{t=1}^{T} P_r(i|X_t, \lambda) \tag{7.6}$$

Mean:

$$\overline{\mu_i} = \frac{\sum_{t=1}^{T} P_r(i|X_t, \lambda) X_t}{\sum_{t=1}^{T} P_r(i|X_t, \lambda)} \tag{7.7}$$

Variance:

$$\overline{\sigma_i^2} = \frac{\sum_{t=1}^{T} P_r(i|X_t, \lambda) x^2_t}{\sum_{t=1}^{T} P_r(i|X_t, \lambda)} \tag{7.8}$$

Here, σ_i^2, x_t, and μ_i refer to the arbitrary elements of the vectors σ_i^2, x_t, and μ_i, respectively.

Expectation Maximization Algorithm

In general, EM iterates through two steps to obtain estimates. The first step is an expectation (E) step, in which missing values are filled in with a guess—that is, an estimate of the missing value given the observed value in the data. The second step is a maximization (M) step, in which the completed data from the E step are processed using ML estimation as though they were complete data; then, the mean and the covariance estimates are updated. Using the newly updated mean and variance matrix, the E step is repeated to find new estimates of the missing values.

The two steps, the E step and the M step, are repeated until the maximum change in the estimates from one iteration to the next does not exceed a convergence criterion. The result of this process is a mean vector and covariance matrix that uses all available information. In other studies [20, 28–33], the EM estimates of the mean vector and covariance matrix can then be used in multivariate analyses to obtain estimates of the model parameters and standard errors, to test a hypothesis, and to score or predict values for observations using the selected model.

Input:

Observed Image in a vector $X_j, j = 1, 2, \ldots, n$ and $i \in \{1, 2, \ldots, k\}$ labels set.

Initialize:

$$\theta^{(0)} = \left(p_1^{(0)}, \ldots, p_k^{(0)}, p_k^{(0)}, \mu_1^{(0)}, \ldots, \mu_k^{(0)}, \sigma_1^{2(0)}, \ldots \sigma_k^{2(0)} \right) \tag{7.9}$$

E-Step:

$$p_{ij}^{(r+1)} = p^{(r+1)} \left(i | X_j \right) \frac{p_i^{(r)} N \left(x_j | \mu_i^{(r)}, \sigma_i^{2(r)} \right)}{f \left(x_j \right)} \tag{7.10}$$

M-Step:

$$\widehat{p}_i^{(r+1)} = \frac{1}{n} \sum_{j=1}^{n} p_{ij}^{(r)} \tag{7.11}$$

$$\mu_i^{(r+1)} = \frac{\sum_{j=1}^{n} p_{ij}^{(r+1)}}{n \widehat{p}_i^{(r+1)}} x_j \tag{7.12}$$

$$\sigma_i^{2(r+1)} = \frac{\sum_{j=1}^{n} p_{ij}^{(r+1)}}{n \widehat{p}_i^{(r+1)}} \left(x_j - \widehat{\mu}_i^{(r+1)} \right) \tag{7.13}$$

Iterate Eqs. (7.12) and (7.13) until a specific error is reached (i.e., $\sum_i e_i^2 < \varepsilon$).

Compute:

$$P_{1j} = \text{ArgMax}_i P_{ij}^{(\text{final})}, \ j = 1, 2, \ldots, n \tag{7.14}$$

In every iteration, the corresponding values of the number of classes, mean, probability, and variance are calculated and used in the iteration.

Support Vector Machine

The SVM was first proposed by Vapnik and has since attracted a high degree of interest in the ML research community. In recent studies, it has been reported that SVMs generally are capable of delivering greater classification accuracy than other data classification algorithms. SVM is a binary classifier based on supervised learning, which performs better than other classifiers. SVM classifies between two classes by constructing a hyperplane in a high-dimensional feature space that can be used for classification. The hyperplane can be represented by the following equation:

$$w.x + b = 0 \qquad (7.15)$$

Here, w is the weight vector and is normal to the hyperplane, whereas b is the bias or threshold. Kernel functions are used with an SVM classifier. A kernel function provides the bridge between non-linear to linear [13]. By using the kernel function, it is possible to map the image with low-dimensional data into a high-dimensional feature space where the data points are linearly separable. The radial bias-based kernel function is used in this SVM-based classification.

Results and Discussion

Classifier Performance

Every classification result could have an error rate and, on occasion, will either fail to identify an abnormality or will identify an abnormality that is not present. This error rate is generally explained in terms of true and false positives and true and false negatives, as follows:

True positive (TP): The result of the classification is positive in the presence of the clinical abnormality.
True negative (TN): The result of the classification is negative in the absence of the clinical abnormality.
False positive (FP): The result of the classification is positive in the absence of the clinical abnormality.
False negative (FN): The result of the classification is negative in the presence of the clinical abnormality.

The following section describes the contingency table, which defines the various terms used to describe the clinical efficiency of a classification based on the previously described terms and the following:

Sensitivity = TP/(TP + FN)*100%
Specificity = TN/(TN + FP)*100%
Accuracy = (TP + TN)/(TP + TN + FP + FN)*100% are used to measure the performance of the classifier.

Contingency Table

In Table 7.1, the features or parameters were extracted using the GMM-based technique. Here, they are used as the inputs for both the GMM-based classifier and the SVM-classifier. Their results vary for each classifier; however, there is not much variation in the classification performance. Here, the radial basis function (RBF) kernel is used in the SVM-based classifier; in the GMM, the K-nearest neighbor method of classification is used. In the SVM classifier, the other kernel-based functions are also compared based on the results of other studies. With this, a Gaussian plot can also be obtained for the input images, which were both normal and abnormal.

In Fig. 7.2a–c, the brain images were taken from the aforementioned dataset, with the images captured by both CT and MRI. This gives the spatial normalization to the montreal neurological institute (MNI) space and to the MRI image with the contrast agents (Fig. 7.2b). Figure 7.2d is a mammogram image that was also used for the training of both the GMM and SVM classifiers.

In Fig. 7.3a–c, the brain images are a set of abnormal images captured through both CT and MRI. They show different brain abnormalities, which are mostly malignant. The mammogram image in Fig. 7.3d shows a tumor; it was also used for the training of both the GMM and SVM classifiers. These images, when trained, also were identified by the classifiers when they were used for testing. The results of the images that were checked after the training were performed using a Matlab simulation environment.

Figure 7.4 shows the Gaussian classifier results for one of the abnormal images from the aforementioned data set that was used in testing. The medical image with

Table 7.1 Contingency table for the classifiers' performance

| | Predicted group | |
Actual group	Normal	Abnormal
Normal	True negative	False positive
Abnormal	False negative	True positive

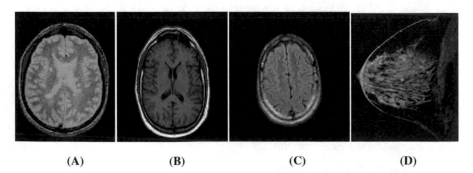

(A)　　　　(B)　　　　(C)　　　　(D)

Fig. 7.2 Normal images for training the classifiers. (**a, b, c**) Brain images. (**d**) Mammogram image

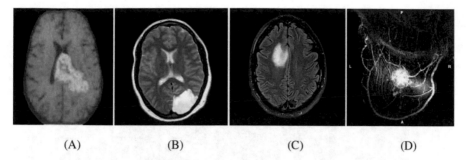

Fig. 7.3 Abnormal images for training the classifiers. (**a, b, c**) Abnormal brain images. (**d**) Abnormal mammogram image

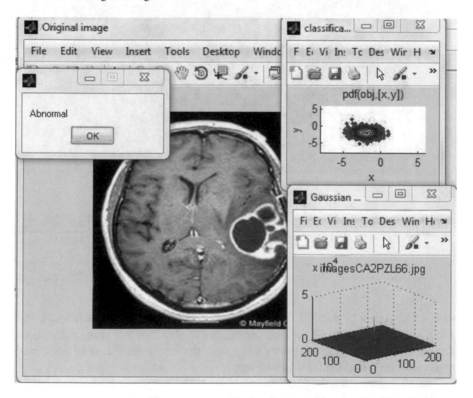

Fig. 7.4 Gaussian mixture model classifier result

the abnormal condition was used for both training and testing of the classifier. The classifier was able to identify the image with the abnormal conditions. Figure 7.4 shows the histogram and maximum likelihood clustering of the pixels because it results in the exact classification of the images.

The result of the brain image used in the dataset for testing by the SVM classifier is shown in Fig. 7.5. The SVM classifier was able to identify the abnormal images

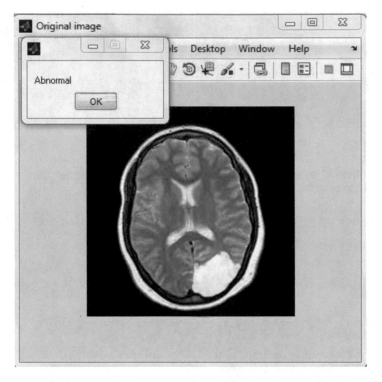

Fig. 7.5 Support vector machine classifier result

Table 7.2 Performance of the classifiers

Classifier	Accuracy (%)
GMM	94.3214
SVM-kernel	96.78
KNN classifier	95.674

with a high value of the cost function, given in the kernel classification as $k = 8$ and $k = 16$. The convergence rate for the SVM classifier has some differences; however, the images were correctly identified, similar to the abnormal ones, with higher accuracy than the GMM classifier.

Table 7.2 shows that the performance of the classifiers was equally good, but the GMM-based model has lower accuracy (94.67%) than the SVM (96.78%) for large dataset values, while still having more features. When trained with a large dataset and the SVM classifier, it worked well in both cases.

Conclusion

The parameters were estimated using MLE based on EM. The GMM model was used as a classifier along with the SVM classifier. Both the GMM and SVM classifiers were trained and tested using the images from the dataset. The MLE had the maximum cluster formation for the given parameters. In the SVM classifier, kernel classification with $k = 8$ and $K = 16$ was used as the given cost function; they performed well when compared to the GMM-based classifier. However, for different cost functions, SVM had a different timing of the convergence rate in the classification but a lower accuracy of 96.78% compared with the GMM-based classifier. The results obtained through these ML-based parameter estimation techniques should be useful in the future to maximize the efficiency of classifications using SVM-based methods.

References

1. Al-Turjman, F., & Alturjman, S. (2018). Context-sensitive access in industrial Internet of things (IIoT) healthcare applications. *IEEE Transactions on Industrial Informatics, 14*(6), 2736–2744.
2. Al-Turjman, F. (2019). 5G-enabled devices and smart-spaces in social-IoT: An overview. *Future Generation Computer Systems, 92*(1), 732–744.
3. Al-Turjman, F. (2018). Modelling green femtocells in smart-grids. *Mobile Networks and Applications, 23*(4), 940–955.
4. Al-Turjman, F. (2018). Mobile couriers' selection for the smart-grid in smart cities' pervasive sensing. *Future Generation Computer Systems, 82*(1), 327–341.
5. Rose, D. C., Arel, I., Karnowski, T. P., & Paquit, V. C. (2010). Applying deep-layered clustering to mammography image analytics. In *2010 Biomedical Sciences and Engineering Conference* (pp. 1–4).
6. Kuang, D., & He, L. (2014). Classification on ADHD with deep learning. In *2014 International Conference on Cloud Computing and Big Data* (pp. 27–32).
7. Bougacha, A., Boughariou, J., Slima, M. B., Hamida, A. B., Mahfoudh, K. B., Kammoun, O., et al. (2018). Comparative study of supervised and unsupervised classification methods: Application to automatic MRI glioma brain tumors segmentation. In *4th International Conference on Advanced Technologies for Signal and Image Processing –ATSIP-2018*, March 21–24, 2018, Sousse, Tunisia.
8. Li, M., & Zhou, Z.-H. (2007). Improve computer-aided diagnosis with machine learning techniques using undiagnosed samples. *IEEE Transactions on Systems, Man, and Cybernetics—Part A: Systems and Humans, 37*(6), 1088–1098.
9. Elguebaly, T., & Bouguila, N. (2011). Bayesian learning of finite generalized Gaussian mixture models on images. *Signal Processing, 91*, 80–820.
10. Yang, S., & Shen, X. (2018). LEEM: Lean elastic EM for Gaussian mixture model via bounds-based filtering. In *2018 IEEE International Conference on Data Mining*.
11. Ramos-Llordén, G., Arnold, J., Van Steenkiste, G., Jeurissen, B., Vanhevel, F., Van Audekerke, J., et al. (2017). A unified Maximum Likelihood framework for simultaneous motion and T1 estimation in quantitative MR T1 mapping. *IEEE Transactions on Medical Imaging, 36*(2), 433–446.
12. Zhang, Z., Zhu, H., & Tao, X. (2017). Image recognition with missing-features based on gaussian mixture model and graph constrained nonnegative matrix factorization. In *Engineering*

in Medicine and Biology Society (EMBC), 2017 39th Annual International Conference of the IEEE, 978-1-5090-2809.

13. Wang, D., Li, Z., Cao, L., Balas, V. E., Dey, N., Ashour, A. S., et al. (2017). Image fusion incorporating parameter estimation optimized Gaussian mixture model and fuzzy weighted evaluation system: A case study in time-series plantar pressure dataset. *IEEE Sensors Journal, 17*(5), 1407–1420.

14. Dutta, A., Ma, O., Toledo, M., Buman, M. P., & Bliss, D. W. (2016). Comparing Gaussian mixture model and hidden Markov model to classify unique physical activities from accelerometer sensor data. In *2016 15th IEEE International Conference on Machine Learning and Applications.*

15. Ji, Z., Xia, Y., Sun, Q., Chen, Q., Xia, D., & Feng, D. D. (2012). Fuzzy local Gaussian mixture model for brain MR image segmentation. *IEEE Transactions on Information Technology in Biomedicine, 16*(3), 339–347.

16. Ihsani, A., & Farncombe, T. (2016). A kernel density estimator-based maximum a posteriori image reconstruction method for dynamic emission tomography imaging. *IEEE Transactions on Image Processing, 25*(5), 2233–2248.

17. Bogunovic, H., Pozo, J. M., Cárdenes, R., San Román, L., & Frangi, A. F. (2013). Anatomical labeling of the circle of Willis using maximum a posteriori probability estimation. *IEEE Transactions on Medical Imaging, 32*(9), 1587–1599.

18. Li, W., Ogunbona, P., de Silva, C., & Attikiouze, Y. (2011). Semi-supervised maximum a posteriori probability segmentation of brain tissues from dual-echo magnetic resonance scans using incomplete training data. *IET Image Processing, 5*(3), 222–232. https://doi.org/10.1049/iet-ipr.2009.0082

19. Crimi, A., Lillholm, M., Nielsen, M., Ghosh, A., de Bruijne, M., Dam, E. B., et al. (2011). Maximum a posteriori estimation of linear shape variation with application to vertebra and cartilage modeling. *IEEE Transactions on Medical Imaging, 30*(8), 1514–1526.

20. De Pierro, A. R., & Yamagishi, M. E. B. (2001). Fast EM-like methods for maximum a posteriori estimates in emission tomography. *IEEE Transactions on Medical Imaging, 20*(4), 280–288.

21. Mihlin, A., & Levin, C. S. (2017). An expectation maximization method for joint estimation of emission activity distribution and photon attenuation map in PET. *IEEE Transactions on Medical Imaging, 36*(1), 214–224.

22. Vapnik, V. (1998). *Statistical learning theory*. New York: Wiley.

23. Scholkopf, B., Kah-Kay, S., Burges, C. J., Girosi, F., Niyogi, P., Poggio, T., et al. (1997). Comparing support vector machines with Gaussian kernels to radial basis function classifiers. *IEEE Transactions on Signal Processing, 45*, 2758–2765.

24. El-Naqa, I., Yang, Y., Wernick, M. N., Galatsanos, N. P., & Nishikawa, R. M. (2002). A support vector machine approach for detection of microcalcifications. *IEEE Transactions on Medical Imaging, 21*(12), 1552–1563.

25. Pontil, M., & Verri, A. (1998). Support vector machines for 3-D object recognition. *IEEE Transactions on Pattern Analysis and Machine Intelligence, 20*, 637–646.

26. Li, W., Lu, Z., Feng, Q., & Chen, W. (2010). Meticulous classification using support vector machine for brain images retrieval. In *2010 International Conference of Medical Image Analysis and Clinical Application*. 978-1-4244-8012-8/10.

27. Said, S., Hajri, H., Bombrun, L., & Vemuri, B. C. (2018). Gaussian distributions on Riemannian symmetric spaces: Statistical learning with structured covariance matrices. *IEEE Transactions on Information Theory, 64*(2), 752–772.

28. Liang, Z., & Wang, S. (2009). An EM approach to MAP solution of segmenting tissue mixtures: A numerical analysis. *IEEE Transactions on Medical Imaging, 28*(2), 297–310.

29. You, J., Wang, J., & Liang, Z. (2007). Range condition and ML-EM checkerboard artifacts. *IEEE Transactions on Nuclear Science, 54*(5), 1696–1702.

30. Reader, A. J., Manavaki, R., Zhao, S., Julyan, P. J., Hastings, D. L., & Zweit, J. (2002). Accelerated list-mode EM algorithm. *IEEE Transactions on Nuclear Science, 49*(1), 42–49.

31. Parra, L., & Barrett, H. H. (1998). List-mode likelihood: EM algorithm and image quality estimation demonstrated on 2-D PET. *IEEE Transactions on Medical Imaging, 17*(2), 228–235.
32. Browne, J., & De Pierro, A. R. (1996). A row-action alternative to the EM algorithm for maximizing likelihoods in emission tomography. *IEEE Transactions on Medical Imaging, 15*(5), 687–699.
33. Liu, J., Ku, Y. B., & Leung, S. (2012). Expectation-maximization algorithm with total variation regularization for vector-valued image segmentation. *Journal of Visual Communication and Image Representation, 23*, 1234–1244.

Chapter 8
Open Source Tools for Machine Learning with Big Data in Smart Cities

Umit Deniz Ulusar, Deniz Gul Ozcan, and Fadi Al-Turjman

Introduction

The idea of storing large volume of data for future processing has been done for decades, while "big data" concept gained momentum in the early 2000s due to the technological advancements that improved Internet access speed, decreased cost of storage, and increased computational power. The big data concept differs from storing large amounts of data in the sense that the speed of data feed has increased to such an extent that it qualifies as a new data source. Also, the data acquisition frequency has increased to a level that new algorithms and techniques for in-depth level of analytics are required.

Growing traffic, increasing population and public safety are major problems of developing cities. Many cities face social and environmental sustainability challenges such as pollution and environmental deterioration [1]. Some have problems of even providing services such as water, sanitation, sewage disposal, and garbage collection. One challenging application area of big data analytics and machine learning that has huge potential to enhance our lives is smart cities.

A smart city infrastructure provides interoperable services between citizens, businesses, and governmental organizations to achieve efficient and sustainable cities. Designing an architecture for the smart city is hence a very complex task, mainly because of the large diversity of devices, communication technologies, and number of services. Constantly changing nature of smart cities and necessity of real-time decision-making require adaptable algorithms and machine learning

U. D. Ulusar (✉) · D. G. Ozcan
Computer Engineering Department, Akdeniz University, Antalya, Turkey
e-mail: umitulusar@akdeniz.edu.tr

F. Al-Turjman
Department of Computer Engineering, Antalya Bilim University, Antalya, Turkey

© Springer Nature Switzerland AG 2020
F. Al-Turjman (ed.), *Smart Cities Performability, Cognition, & Security*,
EAI/Springer Innovations in Communication and Computing,
https://doi.org/10.1007/978-3-030-14718-1_8

techniques. With an enormous growth rate in daily produced data, the challenge is how to efficiently process and learn from it [2]. On the one hand, big data provides great opportunities for researchers to develop algorithms that can extract underlying patterns and to build predictive models on the other hand it creates challenges such as scalability [3].

This chapter provides definitions and explains available tools, libraries, and engines that can be used for big data processing based on criteria such as extensibility, scalability, ease of use, and availability. Also, it provides a comprehensive review of open source tools along with an analysis of the advantages and drawbacks of each technology (i.e., MapReduce, Spark, Flink, Storm, and H_2O) and library (i.e., Mahout, MLlib, SAMOA).

Big Data

Big data can be stated as a huge volume of data that cannot be stored and processed using the traditional approaches. As the cities became more crowded and the advent of ideas such as Internet of Things, deploying sensors all around the cities for intelligent decision-making became a natural step [4, 5]. Due to the living nature of the cities, data may contain valuable information, which needs to be processed in a short span of time. The frameworks discussed in this chapter are able to effectively process data of varying sizes and complexities; hence, they are designed especially for big data. Keeping this in mind, it is important to understand what big data exactly is and how to define it in total of six dimensions commonly called the six V's:

- *Volume* refers to the amount of data that is generated. It encompasses the available data that are out there and need to be assessed for relevance.
- *Velocity* indicates the speed at which data are being generated. Data can be generated and may require to be processed in real-time. Also, data source can be online or offline. As a result, data processing can be classified as batch and stream processing [6]. Batch processing typically works on stored data while stream processing aims to analyze the data in real time as it is generated. Velocity also signifies the rate of change of data and is especially important for stream processing.
- *Variety* refers to the issue of data being in incompatible formats and disparate. It may take significant amount of time to preprocess data that comes in from different sources and in many forms. Data can be structured into a model with predefined columns, data types and so on, whereas unstructured data such as documents, emails, social media text messages, and videos may not have a defined form. The unstructured data requires advanced techniques for storage, data mining, and analysis [6].
- *Veracity* refers to the uncertainty of data. Uncertainty can be in the form of bias, noise, and abnormality [7]. It may be because of poor data quality. Identifying the relevance of data and ensuring data cleansing is required to only store valuable

Table 8.1 Six V's and application of big data in real world situations

Ref.	Volume	Velocity	Variety	Veracity	Variability	Value
Smart city [8, 9]	✓	✓	✓	✓	✓	✓
EHealth [10, 11]	✓	✓	✓			✓
Sensor data [8, 9, 12]	✓	✓	✓			
Risk analysis [13, 14]	✓	✓	✓	✓	✓	✓
Social media [15, 16]	✓	✓			✓	
Financial data [17]	✓	✓		✓		✓

The big data is in extended use in various fields from smart city to financial data

parts and dispose the rest. The main challenge while streaming high-velocity data is the limited time to verify that the data is suitable, can be used for the intended purpose and applicable to the analytic model.

- *Variability* dimension of big data derives from the lack of consistency or fixed patterns in data. It is different from variety in the sense that variability refers to establishing if the contextualizing structure of the data is regular and dependable even in conditions of high level of unpredictability. If the meaning and understanding of data keeps on changing, it will have a huge impact on analysis and attempts to identify patterns [7].
- *Value* deals with the worthiness of data to store and invest in infrastructure, either on premises or in the cloud. It refers to aim, business outcome, or scenario that the solution has to address. Sometimes, data processing needs to also consider ethical and privacy issues.

In Table 8.1, we overview the use of these six V's in Big Data projects. In spite of the importance of these six V's to any big data project, it was not fully considered in all related projects so far, except for the smart city project. This hinders the tight relationship between the big data and the smart city projects which satisfy the core six V features uniquely.

Machine Learning in Big Data

Machine learning is another component in the cloud-based big data paradigm. Many edge devices integrated at the edge of the cloud are added to the smart city architecture for more efficient implementation. The vast amount of information collected from the end-users are valuable for researchers and the smart city operators as well. And, hence, an advanced analytics on smart city applications is needed, where a combination of machine learning algorithms and big data mining techniques is applied. By exploiting the emerging smart city collected data, researchers can develop data-driven solutions for the most pressing issues, such as electricity demands prediction, residential photovoltaic detection, charging

Table 8.2 Types of machine learning algorithms

Supervised learning	Unsupervised learning	Semi-supervised
Bayesian networks	SOM (self-organizing map)	Recommendation systems
Support vector machines	K-means clustering	Reward systems
Decision trees	Hierarchical clustering	
Neural networks	Topic modeling	

demands and points of interest determination, and the time-variant load management problem.

Machine learning deals with the collecting city information in order to provide the cloud-based big data paradigm the ability to learn from its history as humans. It provides information about the properties of the collected data, allowing it to make predictions about other data it may occur in the future. In terms of feedback provided to the algorithms, machine learning is categorized into three: supervised, unsupervised, and semi-supervised learning algorithms. Table 8.2 lists some of the known algorithms in each category.

Supervised Learning Algorithms

These algorithms use training data to generate a function that maps the inputs to desired outputs (also called labels). For example, in a classification problem, the system looks at sample data and uses it to derive a function that maps input data into different classes. Artificial neural networks, radial basis function networks, and decision trees are forms of the supervised learning [18].

Unsupervised Learning Algorithms

This set of algorithms work without previously labeled data. The main purpose of these algorithms is to find the common patterns in previously unseen data. Clustering is the most popular form of unsupervised learning. Hidden Markov models and self-organizing maps are other forms of unsupervised learning [18].

Semi-supervised Learning Algorithms

As the name indicates, these algorithms combine labeled and unlabeled data to generate an appropriate mapping function or classifier. Several studies have proven that using a combination of supervised and unsupervised techniques instead of a single type can lead to much better results [18].

In Table 8.2, the different ML techniques are overviewed and classified according to the aforementioned categories.

Data Availability

Data availability is another issue that effects the type of learning process. Big data can be served as clusters of data or can be real time, and tools need to adapt the velocity of data. Machine learning algorithms are categorized in terms of availability of data into two: batch learning and online learning.

Batch Learning

In the batch learning, entire training data is provided to the system at once. Most of the time data is assumed to be independent of each other and can be divided and processed by the clusters of machines. While processing, clusters do not need to be aware of each other and can process independently which makes the parallel processing simpler.

Online Learning

In the online learning process, data is generated and processed almost instantaneously. Changes in the structure of data, variety and velocity create challenges for the learning systems and needs to be addressed with advanced adaptive algorithms. Typically, developing and deploying a learning system takes time but most of this time is spent on understanding and preprocessing data. Efficient learning can only be possible when usable and valuable data is available. Some issues such as data redundancy, inconsistency, noise, heterogeneity, transformation, labeling for (semi-)supervised learning, data imbalance, and feature selection need to be addressed during data preprocessing stage [19].

- *Data redundancy* and duplication means that at least more than one instances of data represent the same value. Redundancy does not create additional value and creates problems for techniques such as pairwise similarity comparison.
- *Data noise* indicates the parts of the data that needs to be cleaned from missing and incorrect values. Data sparsity and outliers create noise in machine learning models. Manual or human-wise methods are not scalable and inefficient. Some techniques such as replacement of values by the mean decreases the variability and the value of the data.

- *Data heterogeneity* means different data types, different file/data formats, and variability among samples.
- *Data discretization* is the process of converting quantitative data to qualitative data. This process is required/beneficial for some algorithms like Naive Bayes and Decision Trees. Standard discretization algorithms can be paralyzed to cope with big data.
- *Data labeling* is required for supervised learning, semi-supervised learning, transfer learning, and active learning. Online crowd-generated repositories can usually be the source for free annotated data. Typically, experts are used for data labeling in areas like image processing.
- *Imbalanced data* is common for the cases of rare events such as credit card fraud detection. Special care is required for learning from imbalanced data and typically data sampling is performed. When data size increases, sampling has to be addressed using parallel data sampling techniques.
- *Feature selection* is the identification of the properties of data which are more important than others. Feature engineering requires prior domain knowledge and feature selection process is labor intensive.

Open Source Tools for Big Data

There is a great interest in academia, government and private institutions for big data analytics and in the recent years, new technologies and tools have been developed. These tools and frameworks typically approach the problem of processing very large datasets in two ways; data parallelism in which the data is divided into more manageable pieces and each subset is processed simultaneously, or task parallelism, in which the algorithm is divided into steps that can be performed on the same data subset concurrently.

Many of the solutions on big data utilize increasingly complex workflows which require systems built using a combination of state-of-the-art tools and techniques. The most prominent and used open source tools in big data analytics are projects from the Hadoop Ecosystem. The remainder of this chapter provides detailed information about these projects and discusses how they can be utilized together to build an architecture capable of efficiently processing large size data.

Hadoop Ecosystem

As data grows, the most challenging limitation is the scalability of the architecture. Thus, infrastructure scalability handles the changing needs of a big data analysis application by statically adding or removing resources to meet changing application demands. In most cases, this is handled by scaling up (vertical scaling) which is done by adding more resources to an existing system to reach a desired state of performance and/or scaling out (horizontal scaling) which is achieved by distributed architectures.

Apache Hadoop is an open source platform for processing large datasets which uses scale-out architecture instead of scale-up. Additionally, it provides fault tolerance through software and transparently handles server failures. This means that affordable servers can be used for scaling instead of expensive enterprise level fault-tolerant servers. Second, both processing tasks such as batch or stream processing and data extraction, loading and transformation (ETL) operations can be performed. Third, instead of moving data between clusters of computers, moving the code to process data is more efficient and faster which is one of the advantages that Hadoop provides. Finally, because developing applications for a single computer is easier and more manageable than developing distributed applications, Hadoop provides an easy-to-code and manageable framework that keeps developers or analysts away from the complexities of writing code for distributed systems. The ecosystem is composed of four modules:

- *Hadoop Distributed File System (HDFS)*: File system that sits at the bottom of Hadoop architecture made up of data and name nodes with a built-in fault tolerance by keeping copies of nodes in each other.
- *MapReduce*: Data processing engine with two parts, mapping raw data into key–value pairs and processing, and reducing by combining and summarizing the results in parallel.
- *YARN (Yet Another Resource Negotiator)*: Resource manager which allows separation between infrastructure and the programming model.
- *Common*: A set of common utilities like compression codecs, I/O utilities, and error detection.

The general structure of Hadoop ecosystem consists of three layers: storage, processing, and management.

Storage Layer

This layer includes Hadoop Distributed File System (HDFS) and provides a distributed architecture for data storage. HDFS has two main components which are NameNode and DataNode that basically work using master and slave architecture [20]. NameNode is the brain of HDFS system and responsible for propagating data across other nodes. It keeps and handles the metadata of the blocks such as name, size and block count. Due to its unique nature of responsibility, NameNode is considered a single point of failure. DataNodes represent devices used for storing data blocks of the system. With default settings, for availability and fault tolerance, each data block is copied into three different DataNodes. NameNode is also responsible for monitoring the health of DataNodes using a method called heartbeat. DataNode sends a heartbeat at predefined intervals to notify that it is alive. NameNode also acts as a load balancer.

Storage Layer not only consists of HDFS but also can contain non-relational databases, commonly called NoSQL (Not only SQL). Those databases also sit on storage layer of Hadoop ecosystem. Using some open source tools like Sqoop or SQL like query languages (It feels like SQL, but it is not exactly a query language.) data can be retrieved into Hadoop file system. These databases support nested, semi-structured, and unstructured data which we commonly face with as the data rapidly grows. In terms of simplicity, we will not delve into more details; knowing that Hadoop supports and provides open source tools for the integration of NoSQL databases into HDFS is enough for this chapter.

Processing Layer

Analysis of big data is achieved at the processing layer. YARN creates an environment where one or more processing engines can run on Hadoop data cluster simultaneously. Before YARN was introduced, there were Job Trackers and Task Trackers in Hadoop 1.x architecture. Those were good at managing resources but were creating bottlenecks hence inefficient. To overcome performance bottlenecks, Hadoop has introduced YARN instead of tracker models in Hadoop 2.x. YARN divides the functionality of trackers into four services, namely, Resource Manager, Node Manager, Container, and Application Manager.

Processing models used by data processing engines are categorized as either batch or streaming. Batch processing is used on large datasets and the output is written to a file or database when execution is completed. There is no real-time work in batch processing whereas in stream processing, data is processed as it arrives at the system. This is good for real-time analysis of data but this type of processing requires special techniques to assure that the results are obtained in a meaningful time. The list of keywords used for evaluation of processing engines are as follows.

- *Latency*: Time between the start of a job and the initial results.
- *Throughput*: Amount of work done over a given time period, efficiency.
- *Fault Tolerance*: Failure detection and recovery options.
- *Usability of Engine Itself*: Complexity of installation and configuration, interface language, and programming.
- *Resource Expense*: Financial and time-wise cost.
- *Scalability*: It indicates the ability of the system to adapt the increasing demands of requirements. The scalability of a processing engine is a major factor and tries to answer if there will be a bottleneck when input or cluster sizes grow.

MapReduce

Using the ideas originated from functional programming and Google MapReduce library, Hadoop introduced the MapReduce framework to facilitate parallel processing. The MapReduce model consists of two main operations, Map and Reduce. Map is the process of acquiring data from storage, applying the algorithm and generating results in the form of key–value pairs. Reduce is the process of performing aggregation functions such as summation, multiplication, on the key–value pairs. Some operations such as shuffle and sort are available for data processing in between mapping and reducing.

Similar to HDFS, MapReduce also has a master node, which orchestrates the cluster-wide operations and worker nodes. Whenever a task fails, fault tolerance is achieved by simply re-execution of the task. If a node's process rate is less than other nodes, called stragglers, same tasks are assigned to other nodes.

MapReduce is also used for machine learning, in which the training data set is read entirely to build a learning model. In a batch-oriented workflow, the data is read from the HDFS to the mapper. Mapper produces and writes those key–value pairs to disk to be sorted. After sorting, intermediate data is sent to reducer to train a model. These write and read operations can be inefficient in terms of time and computational resources.

Spark

Apache Spark is also a cluster-computing environment and uses ideas similar to MapReduce model, but improves speed by using in-memory computations. Its response time is significantly faster than MapReduce in processing tasks stored in memory and Hadoop at disk operations. It stores data in memory and provides fault tolerance without replication with abstraction called Resilient Distributed Datasets. RDD can be understood as read-only distributed shared memory. The RDD was extended to include DataFrames. This allows grouping of collection of data by columns hence it can be thought as RDD with schema. Learning process is through in-memory caching of intermediate results. Spark is easy to program and supports integration with Java, Python, Scala, and R programming languages. It supports multiple data sources, including Cassandra, HBase, or any Hadoop data source. Besides its effective features, Spark has some inefficiencies in terms of stream processing and bottlenecks can occur because of data transfer across nodes using network [21].

Spark has different processing components, which are as follows in simple terms:

- *Spark Core*: The main component of Spark. Critical functions like task scheduling take place here.
- *Spark SQL*: Structural data processing unit. It runs both SQL and Hive query language queries for structured data.
- *Spark Streaming*: Stream processing unit for real-time processing of data.
- *MLib*: Machine learning library is designed to execute algorithms like classification, regression, decision trees, random forests, and gradient-boosted trees.
- *GraphX*: It is designed for graph-related analytics.

Storm

Storm is specifically designed to handle streaming data and offers stream processing in real time. It is comprised of spouts (input stream) and bolts (computation logic). Bolts can process the data coming from both spouts and other bolts. Storm uses real-time streaming but also offers micro-batch via its Trident API. Storm is primarily implemented in Closure, initially in Java, and now includes Thrift for cross-language development. Fault tolerance is achieved by a job tracker called Nimbus. Nimbus keeps track of the worker nodes. If Nimbus dies, it is restarted automatically differing from MapReduce's jobtracker method. Storm also supports "Lambda Architecture" which is an approach of breaking downstream processing into three layers: batch, serving, and speed. MapReduce and Storm can run jobs simultaneously that can be used to process both real-time and historical data. Storm does not include a machine learning library, but SAMOA (big data mining platform) offers implementations for typical classification and clustering algorithms.

Flink

Apache Flink is a framework for processing unbounded and bounded data streams in cluster environments. Unbounded data can be best described as a type of ever-growing, essentially infinite data set. These are often referred to as "streaming data." Bounded data is finite and such as batch data sets. Flink's processing model applies transformations to parallel data, generalizing map and reduce functions, and functions like join, group and iterate. Flink has capability for true batch and stream processing data in real time. It offers APIs for both Java and Scala and scalable. Flink can be both integrated with HDFS and YARN or run independent from Hadoop ecosystem. It offers optimization mechanism like cost-based optimizer and iterative batch as well as streaming options. A machine learning library called Flink ML was also introduced. It also offers adapters for other machine learning libraries such as SAMOA.

H_2O

H_2O differs from other processing engines with its web-based user interface. This interface makes machine learning tasks more understandable and accessible to non-technical users. Also, it offers parallel processing engine, analytics, math and machine learning libraries along with data preprocessing and evaluation tools. Framework supports Java, R, Python, and Scala programming environments. Streaming is provided with Sparkling Water (creative name with the union of names of Spark and H_2O) project which is the integration of Spark into H_2O. It processes data completely in memory using distributed fork/join, a divide and conquer technique, for parallel tasks.

Management Layer

Management layer is responsible for scheduling, monitoring and coordination and provides a user interface. As the tools for storage and processing are used, creating tasks require organization and fine tuning. This high-level organization and user interaction take place in management layer. Some of the tools are as follows.

Oozie

This is a system for running and scheduling Hadoop jobs. It is basically an interactive workflow scheduler and allows to chain stuff together. For example, one can chain together MapReduce, Hive, Pig, Sqoop, and other tasks. A workflow is a directed acyclic graph of actions specified via xml. So one can run actions that do not depend on each other.

Zookeeper

This is a service for coordination and synchronization, and keeps track of information on master and worker nodes, task–worker map, and worker availability. ZooKeeper is a tool that applications can use to recover from partial failures. Zookeeper can handle failures such as weird partitions and clocks drift. It provides support for Java, C, and also has bindings for Perl, Python, and REST clients. Primitive operations of Zookeeper are master election, crash detection and group management.

Hue

Hue offers web interface for Hadoop ecosystem. It features file browser for HDFS and job browser for MapReduce and YARN. Hue can be used to manage interactions with Hive, Pig, Sqoop, Zookeeper, and Oozie, and also offers tools for data visualization.

Machine Learning Toolkits

In this section we list some of the open source machine learning toolkits for big data processing. Each toolkit is examined according to the following parameters:

- *Scalability*: Size and complexity of the data currently available and will reach in the future.
- *Speed*: Frequency of the updates.

- *Coverage*: Range of options contained. More tools can be unnecessary and difficult to set up.
- *Usability*: Initial setup, maintenance, programming languages, UI, documentation and user community.
- *Extensibility*: How well to build blocks for new platforms.

Mahout

It offers wide selection of robust algorithms. Mahout is good for batch processing (not streaming). There is a lack of active user community and documentation. It is commonly claimed to be difficult to set up an existing Hadoop cluster. Configuration problems may occur. Algorithms focus on classification, clustering and collaborative filtering. Extensibility is good but strong java knowledge is required. Mahout is best known for collaborative filtering (recommendation engines) offers similarity measures like Pearson correlation, Euclidean distance, Cosine similarity, Tanimoto coefficient, log-likelihood, and others.

MLLib

MLLib covers the same range of learning categories as Mahout but also adds regression models. There are some additional tools like dimensionality reduction, feature extraction and transformation, optimization, and basic statistics. It is relatively easy to set up and run. User community is not that active.

H_2O

H_2O is also considered a product for machine learning rather than a project. It offers graphical user interface (GUI), and numerous tools for deep neural networks. H_2O has a good documentation and offers classification, clustering, generalized linear models, statistical analysis, ensembles, optimization tools, and data preprocessing options.

Samoa

Scalable Advanced Massive Online Analysis (SAMOA) is a platform for machine learning for streaming data. Feedbacks can occur in real time. SAMOA is a flexible environment and can run additional stream processing engines like Storm, S4, and Samza [22]. Supports algorithms for classification, clustering, regression, and frequent pattern mining, along with boosting, and bagging for ensemble creation [6].

Data Movement and Integration Tools

In addition to data processing engines and machine learning libraries, storage layer of Hadoop ecosystem also includes tools for data movement and interaction.

Kafka

Kafta is a general-purpose distributed publish–subscribe messaging system on top of HDFS. It can handle data coming from multiple resources in real-time. It is not only for Hadoop; independently can be used for any kind of messaging implementations. Kafka servers store all incoming messages from publishers for some period of time and publishes them to a stream of data called a topic. Kafka consumers subscribe to one or more topics, and receive data as it gets published. A stream/topic can have many different consumers, all with their own position in the stream maintained. Architecturally, you can think of a Kafka cluster at the center of a producer network. Producers might be individual apps listening for data. They communicate with the thing generating the data and Kafka. Consumers subscribe to the topics and receive data. Connectors—modules for databases, plugins that allow it to receive new messages and store, or actually publish changes into Kafka itself. Stream Processors transform data as it comes in.

Flume

This is another way to stream data into data cluster like Kafka. It is initially intended for collection, aggregation and movement of log data into HDFS. Log traffic can be spiky and can perform some bottlenecks in the cluster. By introducing Flume in the middle, one can create a middleware that will not bring down the entire cluster and get everything to catch up. Flume acts like a buffer between data and cluster.

Sqoop

Sqoop is another tool that is used to import and export data between relational databases and Hadoop ecosystems. This is useful when HDFS is used as an enterprise data warehouse preprocessing engine. The idea behind Sqoop that it leverages map tasks of MapReduce framework.

Hive

Storage layer also includes data integration tools such as Hive, which allows for running standard SQL queries on data stored in the HDFS and NoSQL databases using HiveQL, an extension of ANSI SQL. This is a powerful and simple way to query the system, which then is distributed across MapReduce/TEZ commands, and then runs on top of Hadoop YARN. Metadata for tables and partitions is kept in the Hive Metastore. HIVE provides interactive way of working with big data on a cluster and way easier than writing MapReduce code in Java. It is highly optimized and extensible. Hive is good for online transaction processing and stores data denormalized as flat text files. No record level updates, inserts or delete are allowed because of not existing relational database underneath.

Open Research Issues

Recently, popular technology solutions for various smart cities applications have emerged as a significant advancement for Internet and cloud computing paradigms. Big data is one of the novel cloud paradigms where connected entities become part of the smart city infrastructures, and the advancements in cloud computing make it quite popular where the traditional telecommunication systems facilitate basic communications between the cloud entities. Big data has converged technologies in terms of services, computing, information processing, networking and controlling intelligent technologies. Among the key technologies converged is the huge data processing due to its computing capability and cost effectiveness. Wealth of various approaches have been proposed and designed for considering the collaborative nature of the cloud in the existing literature. However, there are key open research problems still to be solved in the big-data era.

Since, in general, enabling technologies have restricted authentication privileges for mobile users, different strategies are introduced for the extension of user authentication over the cloud-based environments. Commercialization of remote applications, and security issues in big data have gained much attention of the researchers to satisfy the security properties of authentication and key agreement protocols. In general, the development of security protocols is more challenging and should also consider mitigation of the computation and communication cost. Moreover, considering the big data processing and the cloud energy consumption for heterogeneous networks in smart cities is another key challenge. Where energy consumption models, considering the aforementioned smart-city setups are mandatory for more accurate estimations while considering different access types [23]. Furthermore, covering a large area with the optimized deployment of the cloud infrastructure and many overlapping services needs a careful consideration in order to realize an efficient big data paradigm in the smart-city era [9].

Conclusion

As the cities became more crowded and the advent of ideas such as Internet of Things, deploying sensors all around the cities for intelligent decision-making became a natural step. As the amount of data increased significantly, researchers developed new techniques for handling both streaming and stored data originating from different sources.

In this chapter, we see some of the tools for big data analytics and discuss techniques that can be used. There is no single tool or framework that covers all or even the majority of big-data processing tasks; one must consider the trade-offs that exist between usability, performance, and algorithm selection when examining different solutions. So far, the traditional algorithms and tools have mostly helped to solve the data processing requirements, but as the size of data grows, the scalability problem arises.

References

1. Haughton, G., Hunter, C., & Hunter, C. (2004). *Sustainable cities*. London: Routledge.
2. Seetha, H., Murty, M. N., & Tripathy, B. K. (2001). *Modern technologies for big data classification and clustering*. Hershey, PA: IGI Global.
3. Zhou, L., Pan, S., Wang, J., & Vasilakos, A. V. (2017, May). Machine learning on big data: Opportunities and challenges. *Neurocomputing, 237*, 350–361.
4. Ulusar, U. D., Celik, G., Turk, E., Al-Turjman, F., & Guvenc, H. (2019). Practical performability assessment for ZigBee-based sensors in the IoT era. In F. Al-Turjman (Ed.), *Performability in internet of things* (pp. 21–31). Cham: Springer.
5. Ulusar, U. D., Celik, G., & Al-Turjman, F. (2017). Wireless communication aspects in the internet of things: An overview. In *2017 IEEE 42nd Conference on Local Computer Networks Workshops (LCN Workshops)* (pp. 165–169).
6. Landset, S., Khoshgoftaar, T. M., Richter, A. N., & Hasanin, T. (2015, November). A survey of open source tools for machine learning with big data in the Hadoop ecosystem. *Journal of Big Data, 2*(1), 24.
7. Somani, A. K., & Deka, G. C. (2017). *Big data analytics: Tools and technology for effective planning*. New York: Routledge.
8. Al-Turjman, F. (2019). *Intelligence in IoT-enabled smart cities*. Boca Raton, FL: CRC Press.
9. Al-Turjman, F. (2019, March). 5G-enabled devices and smart-spaces in social-IoT: An overview. *Future Generation Computer Systems, 92*, 732–744.
10. Chawla, N. V., & Davis, D. A. (2013, September). Bringing big data to personalized healthcare: A patient-centered framework. *Journal of General Internal Medicine, 28*(3), 660–665.
11. Suciu, G., et al. (2015, September). Big data, internet of things and cloud convergence – An architecture for secure E-health applications. *Journal of Medical Systems, 39*(11), 141.
12. Al-Turjman, F. (2019, March). Cognitive routing protocol for disaster-inspired internet of things. *Future Generation Computer Systems, 92*, 1103–1115.
13. Chan, P. K., Fan, W., Prodromidis, A. L., & Stolfo, S. J. (1999, November). Distributed data mining in credit card fraud detection. *IEEE Intelligent Systems and Their Applications, 14*(6), 67–74.
14. Choi, T.-M., & Lambert, J. H. (2017). Advances in risk analysis with big data. *Risk Analysis, 37*(8), 1435–1442.

15. Tsou, M.-H. (2015, August). Research challenges and opportunities in mapping social media and big data. *Cartography and Geographic Information Science, 42*(sup1), 70–74.
16. Cappella, J. N. (2017). Vectors into the future of mass and interpersonal communication research: Big data, social media, and computational social science. *Human Communication Research, 43*(4), 545–558.
17. Sohangir, S., Wang, D., Pomeranets, A., & Khoshgoftaar, T. M. (2018, January). Big data: Deep learning for financial sentiment analysis. *Journal of Big Data, 5*(1), 3.
18. Al-Turjman, F., Imran, M., & Bakhsh, S. T. (2017). Energy efficiency perspectives of femtocells in internet of things: Recent advances and challenges. *IEEE Access, 5*, 26808–26818.
19. Hassanien, A. E. (2018). *Machine learning paradigms: Theory and application*. London: Springer.
20. Syed, M. F. A. (2018). *Big data architect's handbook*. Birmingham, UK: Packt.
21. Guller, M. (2015). *Big data analytics with spark: A practitioner's guide to using spark for large-scale data processing, machine learning, and graph analytics, and high-velocity data stream processing*. New York: Apress.
22. Swathi, R., & Seshadri, R. (2017). Systematic survey on evolution of machine learning for big data. In *2017 International Conference on Intelligent Computing and Control Systems (ICICCS)* (pp. 204–209).
23. Al-Turjman, F. (2018, July). Fog-based caching in software-defined information-centric networks. *Computers and Electrical Engineering, 69*, 54–67.

Chapter 9
Identity Verification Using Biometrics in Smart-Cities

D. R. Ambika, K. R. Radhika, and D. Seshachalam

Acronyms

FAR	False accept rate
FRR	False error rate
IoT	Internet of Things
LBP	Local binary patterns
LCPR	Lower central periocular region
RAM	Random access memory
ROC	Receiver operating characteristics
ROI	Region of interest
SSIM	Structural similarity index

Introduction

Security plays key role in determining prospects of a smart city. Currently numerous efforts are made by researchers in exploiting technology advances to achieve public safety. Efficient network security models are proposed to address vulnerabilities, threats, attacks, and risks in the Internet of Things (IoT) era [5, 6, 9]. Certain

D. R. Ambika (✉) · D. Seshachalam
Department of Electronics and Communication Engineering, BMS College of Engineering, Bengaluru, India
e-mail: amusatish@gmail.com

K. R. Radhika
Department of Information Science Engineering, BMS College of Engineering, Bengaluru, India
e-mail: rkr.ise@bmsce.ac.in

© Springer Nature Switzerland AG 2020   169
F. Al-Turjman (ed.), *Smart Cities Performability, Cognition, & Security*,
EAI/Springer Innovations in Communication and Computing,
https://doi.org/10.1007/978-3-030-14718-1_9

researchers propose agile frameworks and context-sensitive seamless identity provisioning (CSIP) framework to enhance authentication, confidentiality, and integrity for IoT architectures [7, 8]. Biometrics suggests a smart solution to keep the city safe. In addition to surveillance benefits of biometrics, an identity-recognition app on a mobile device allows officers to click photo of a suspect instantly and verify identity on spot, without any overhead of bringing suspects to station. Such a module, being able to be installed as an app, poses several restrictions. The system demands computational simplicity coupled with the ability to easily match a suspect across several subjects. Manual pre-processing procedures need to be completely eliminated, alleviating any dependence on user inputs. Storage efficiency, ability to scale and store several hundreds of biometric templates and high authentication accuracy despite varied conditions of illumination, expression, pose, scale, distance and the presence of hair, masks, and other occlusions are essential features of such a biometric system. Further, system needs to be resistant to cosmetic modifications, for instance, shaving, shaping, or growing of eyebrows/beard/hair/mustache.

A number of biometric modalities are explored by researchers in the recent years. Face images are one of the most widely used traits due to their discriminative abilities. Non-intrusive acquisition of facial information has promoted increased preference over other reliable traits, for instance, fingerprint. Face biometric systems are currently available, and ample algorithms are developed using both still images and videos. Certain challenges remain to confront face biometric applications. Complete face image capture is largely hindered due to head orientation, masking, social non-acceptance, intervention of hair, cap, and other accessories. Significant deformation is witnessed in different portions of face in expression variant images. Cosmetic modifications significantly impact the potentiality of the face template. Further, a full-face database over a large population insists high computation and memory requirements. In contrast, iris texture is predominantly unique for each individual and requires lesser storage space and computation time. The characteristics remain stable over a large period of time. An important concern of iris-based systems is reliable acquisition for non-cooperative and at-a-distance subjects. Poor quality images significantly impact the performance of the system. Imaging retinal vasculature requires the user to cooperate without blinking or varying the pose. Accordingly, improved image acquisition methodologies and enhanced feature extraction mechanisms are required to address the discussed challenges. A related idea is to explore newer traits that aid the existing systems. Periocular region is one such unique trait for recognition purposes, after preliminary exploration of its feasibility [28, 29]. The region adds value in comparison to many existing biometric solutions. Periocular refers to region around the eye encompassing anatomical features including upper/lower eye fold, eyelids, wrinkles, hair follicles, pores, and skin tags. Eyebrows are also included within the region of interest for certain applications. Periocular biometrics is regarded as an emerging technology towards unconstrained authentication.

Current work proposes a novel authentication technique exploring potential regions of periocular region. The study demonstrates that verification accuracies attained by utilizing only a sub-region of periocular area commensurate with the state-of-the-art performances obtained using complete periocular region. The results suggest that potential portions of periocular region, coupled with the proposed feature extraction framework, contribute significantly towards computational efficiency; and hence its deployment in memory restricted applications.

Highlights of the Current Approach

Potential Benefits of Utilizing Periocular Region as a Useful Biometric Trait

Ease of Acquisition   Vital asset of periocular region is its reliable acquisition over a wide range of distances, contradicting the needs of other biometric traits. Acquiring iris images at large standoff distances results in low texture resolution. Facial images acquired at close quarters cause loss of certain face information. Periocular region features are available for analysis in scenarios where quality of iris images is poor, not containing subtle details or when face images are occluded [15]. The region of interest around eye offers a good trade-off between iris and face biometric templates.

Computational Efficiency   Periocular region caters to the needs of real-time applications. Size of a facial template being large slows down the system for a large database, as big as a nation. On the contrary, periocular region being less than 25% of the entire face fastens the authentication mode. In addition, the trait is useful in assisting face data, costing no extra acquisition effort and storage space [12].

Useful Periocular Features   Periocular region encompasses rich discriminative features. Level-one periocular features include upper or lower eye folds, upper or lower eyelids, wrinkles, and moles. Detailed and fine features including skin textures, pores, and hair follicles form level-two features [9]. Structural attributes of periocular region include curvature of eye boundary, tapering of eye shape towards eye corners, size and location of tear duct, shape and density of eyebrows, depth of eye socket, location and shape of birth marks [10].

Biomechanical Nature of Periocular Region   In accordance with human anatomy, large number of muscles interacts more in the face than any other part of the human body. Muscles from orbicularis oculi, nasalis, procerus, quadratus labii superioris, and levator labii superioris families determine the non-linear distortions due to facial expressions [6]. Useful biomechanical aspects of periocular region that assist in attaining reliable authentication for expression variant images are: (1) sudden local angular distortions rarely occur. Specifically, angles between adjacent regions tend to be preserved (2) strong spatial correlation of displacements is determined by skin elasticity [40].

Potential Periocular Sub-Region

Different regions of periocular area are examined in research using various mathematical tools. A comparatively larger template is more useful for classification, but is accompanied by intense feature extraction schemes and increased storage space [12]. An optimized template incorporating sufficient discriminative features and reduced in size is required for a candid human verification system. Periocular region-based biometric experimentation largely involves inclusion of iris. A few researchers eliminate iris texture by placing a circular mask of neutral color using pupil co-ordinates [17]. An elliptical mask of neutral color is placed in the middle of periocular region as an alternative to eliminate surrounding sclera along with iris texture [33]. Inclusion of eyebrows report non-reliability under varied expression and cosmetic modification although the characteristics provide additional features during matching [18]. Consequence of eliminating eyebrows and sclera region leaves subtle entropy in the upper periocular region. Upper portion effectively constitutes curvatures of upper eye fold and eyelids, as shown in Fig. 9.1b. Lower portion of periocular region, termed as lower central periocular region (LCPR), encompasses more anatomical details. Bulging of lower eye, edges/ridges along the bulge, curvature of eye socket, depth of socket towards either ends of the eye, and foldings on lower eye fold constitute rich density of texture information as shown in Fig. 9.1c. Ancillary factor motivating LCPR region is that the region is less susceptible to occlusion and deformation. The presence of hair mostly occludes eyebrows and upper region of eye, while LCPR survives occlusion and is sufficiently available for processing. Movement of eyeball affects the curvatures of upper eyelid more than the LCPR as seen in Fig. 9.2. The figure depicts periocular and LCPR regions for occlusion due to hair and varied pose images. Variation in expression causes significant deformation in the upper region of periocular area, particularly, texture of eyebrows and upper eye fold. Figure 9.3 shows periocular and LCPR images for varied expressions. The figure illustrates that the distances between upper eye fold and eyebrows vary largely over change in expression. In addition, visible portions of sclera region and eyeball available are different when eyes are opened

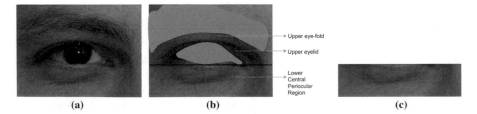

Fig. 9.1 Illustration of texture features in upper and lower periocular region. (**a**) Periocular image. (**b**) Elimination of sclera and eyebrow information

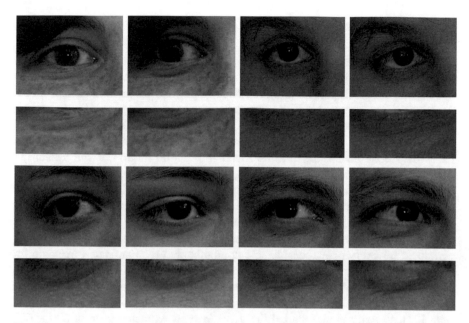

Fig. 9.2 LCPR regions extracted from periocular regions of different subjects exhibiting hair occlusion and variation in pose

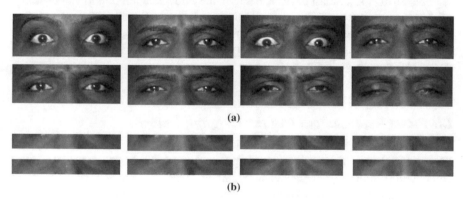

Fig. 9.3 (a) Periocular regions for different expressions. (b) Corresponding LCPR regions

differently. LCPR undergoes subtle or no deformation under varied expression and eyeball movement as witnessed from the figures. In contemplation of the benefits using LCPR and its predominant contribution towards periocular entropy, current experimentation investigates authentication by using LCPR.

Computationally Efficient Variation of LBP

Local binary patterns (LBP) is one imperative approach largely used by periocular researchers for skin texture description. Local descriptors are computed by dividing the region into small patches and the resulting patterns are combined into global description. The approach draws attention particularly due to its ability to capture small appearance details. An additional motivation is that human perception is well adapted to extracting local information as micro-patterns from images [2]. LBPs have been demonstrated to be successful texture descriptor and tolerant to illumination variation [25, 26, 28, 34, 38, 39]. A number of periocular researchers have investigated biometric systems using LBP and its variation or fusion of LBP with other feature extraction schemes [1, 19, 27, 31]. Acceptable performance is attained even under unconstrained environment. Reliable efficacy of LBP for texture description is attained at the cost of increased processing time. LBP features being computed for every pixel, considerable computational requirement is mandate for real-time feature extraction applications. Miguel et al. reported that a processing speed of 22–125 Megapixels per second is required for high definition video resolution (1280×720–1920×1080) using 25–60 frames per second as refresh rates [23]. Vazquez et al. reported that LBP features extracted over an 800×600 pixel picture for a mobile environment using 1 GHz ARM Cortex A8 processor requires 36 ms [36]. Smart devices, being battery powered, an important constraint present is to reduce the computational power. A computationally efficient variation of LBP is introduced in the present chapter by utilizing entropy of a dominant bit-plane of LCPR image. The method does not compromise on the performance, and suits smart device applications.

Bit-Plane Representation of the Original Image

Binary bit-planes of an image are largely adopted for image retrieval, image compression, and steganographic applications. One prominent applicability for texture feature extraction is to apply wavelet transform on individual bit-planes of a grayscale image and acquire multi-resolution binary images. Resulting images are compressed forms with no loss of useful information for many medical applications [21]. The current work adopts bit-plane slicing during feature extraction to determine a single dominant bit-plane that sufficiently abstracts the original image. Local binary patterns are computed over dominant bit-plane of an LCPR image rather than over the input intensity image. Proposition of dominant bit-plane LBP is emanated from the following observations:

1. Crucial textural features of periocular region are edges, ridges, curvatures, eye folding, and wrinkles along with socket depth information. Although an entire intensity image encompasses the textural patterns, certain bit-planes of the image are significant contributors.

2. Certain intermediate bit-planes are more relevant textural descriptors over other bit-planes. In consequence, a dominant plane is sufficient for LBP feature extraction. The resulting procedure is useful to reduce computational complexity without performance degradation.

Feature Extraction Using Dominant Bit-Plane LBP

The proposed framework involves four important stages of computation, as shown in Fig. 9.4. The stages are segmentation of LCPR region, determination of dominant bit-plane, extraction of dominant bit-plane LBP feature vectors, and classification.

Segmentation of LCPR

Primary goal to segment the region of interest is to determine the upper boundary of LCPR. Contour around the eyeball, specifically towards lower eyelid, forms the required upper perimeter. Detection of eyeball region aids in determining the

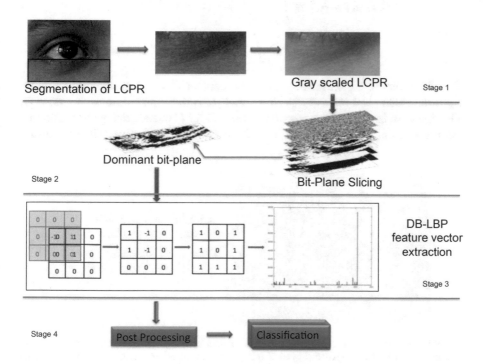

Fig. 9.4 Important stages involved in dominant bit-plane LBP framework

eyeball boundary. Image segmentation techniques, for instance, level set-based segmentation, geodesic active contour segmentations, and shape constrained level set
representations are useful to derive contour boundaries that are parameter free, and
change naturally to the topology [32]. The task of eyeball localization in the present
work is performed by an automated mechanism using active contour segmentation
technique. The appropriateness of using active contour method of segmentation is
due to the fact that the method efficiently segments regions of different intensity
gradients. Periocular region P_{peri}, being formed by two sub-regions, homogeneous
foreground eyeball region and non-homogeneous background region, invokes to
employ Chan–Vese active contour method to localize the eyeball [16].

P_{peri}^i represents piecewise-constant intensity of eyeball region and P_{peri}^o represents the non-homogeneous background region constituting skin texture and
eyelashes as shown in Fig. 9.5. On considering $C_{eyeball}$ as the eyeball contour,
P_{peri} is approximated to P_{peri}^i inside the $C_{eyeball}$ and to P_{peri}^o outside the $C_{eyeball}$.
The model starts with a curve C around the eyeball. The curve moves to the
interior normal based on the image statistics and stops on the boundary of eyeball.
Equation (9.1) is the objective function representing the contour evolution. The basic
idea of Eq. (9.1) is to evolve a contour on the boundary of eyeball region using the
principle of energy minimization.

$$E(C, \mu_e, \mu_b, \lambda_1) = \int_{P_{peri}/C} |P_{peri}(x) - \mu_b|^2 dx$$

$$+ \lambda_1 \int_C |P_{peri}(x) - \mu_e|^2 dx + \lambda_2 \Re(C) \qquad (9.1)$$

P_{peri}/C refers to region inside the current contour C. $\Re(C)$ is regularizing term,
μ_e is the mean pixel intensity inside C, and μ_b is mean pixel intensity outside C.
The first term is > 0 and the second term is ≈ 0 for C outside the eyeball. The first
term is ≈ 0 and the second term is > 0 for the C inside the eyeball. As a result,

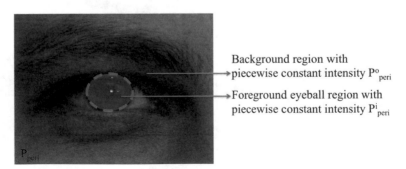

Fig. 9.5 Illustration of two sub-regions of input periocular image each with different piecewise-
constant intensities: homogeneous eyeball region surrounded by non-homogeneous background

fitting term is minimized for $C = C_{eyeball}$, i.e., the curve C is on the boundary of the eyeball.

Initial contour plays a crucial role in eyeball detection procedure. Positioning the contour initially within eye region ensures reliable localization. Initial contour based on Harris corner detector is used to confine evolution within eye region. The detector is used to identify the brightest point present in the proximity of eyeball center. Co-ordinates of the detected point are used to position initial contour towards the center of eyeball region. Figure 9.6 shows successful detection of eyeball center and subsequent positioning of initial contour using corner detection algorithm. Evolution of contour at steps of 20 iterations is illustrated in Fig. 9.7. On experimentation it is observed that localization is achieved between 150 and 200

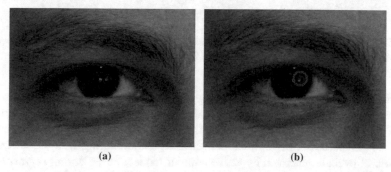

(a)   (b)

Fig. 9.6 (a) Co-ordinates of eyeball center determined by the corner detector. (b) Positioning initial contour in the vicinity of eyeball region

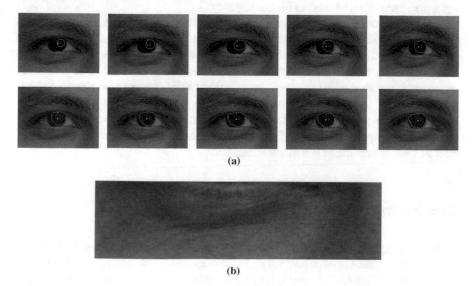

(a)

(b)

Fig. 9.7 (a) Evolution of active contour during LCPR segmentation at steps of 20 iterations. (b) Segmented LCPR region after iris localization using 200 iterations

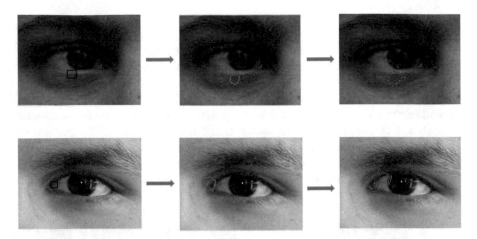

Fig. 9.8 False localizations resulting due to error in initial contours through wrong determination of eyeball centers

iterations. Subsequent eyeball localization for all periocular images in experimental datasets is achieved using 200 iterations. Occlusions from eye lashes and eye brows result into formation of small contours during contour evolution. The extraneous contours are discarded based on their sizes and position during cropping of LCPR. Contour of eyeball generated by active contour model is used during segmentation to determine the upper bound of LCPR. Lower bound is placed at lower perimeter of input periocular images. Left and right bounds are placed at left perimeter and right perimeter of input images for consistency.

Erroneous initial contours contributed by errors in determining eyeball centers result in false localizations. Figure 9.8 shows results of false localizations. An experimental validation of automatic segmentation approach is performed by segmenting LCPR manually from input periocular images. Comparison of automatic segmentation with manual approach reveals that LCPR regions resulting from both the approaches are similar. 99% of automatically segmented ROI corresponds to manually segmented images. Figure 9.9 shows samples of LCPR images segmented using both the approaches. 1% of mismatch is accounted by incorrect detection of eyeball centers causing errors in initial contours.

Construction of Dominant Bit-Plane

A digital image is composed of pixels, with each pixel representing a particular quantization level of an intensity value. Binary equivalent of each pixel is composed of a bit-word (8-bit size). Individual bits contribute different levels of intensities. A set of bits corresponding to a given bit position of each binary number representing

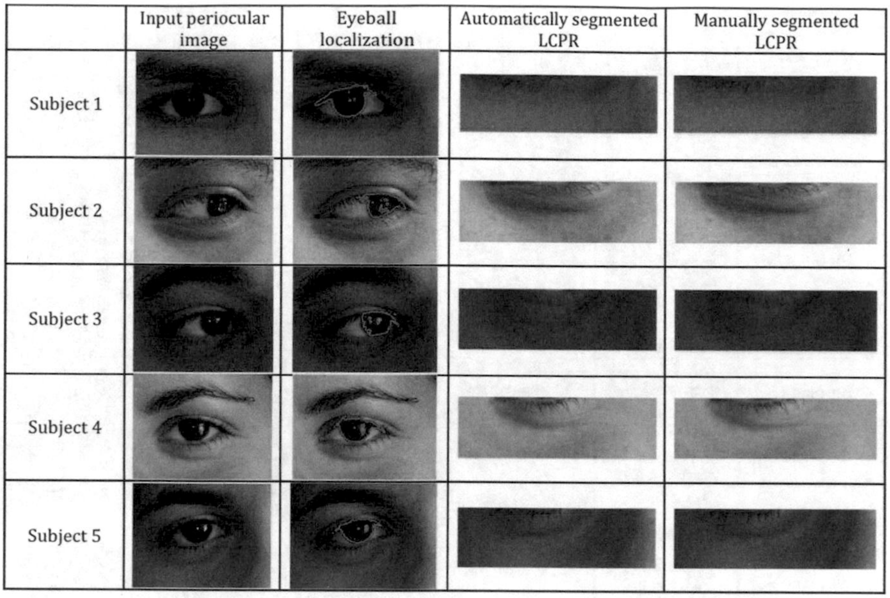

	Input periocular image	Eyeball localization	Automatically segmented LCPR	Manually segmented LCPR
Subject 1				
Subject 2				
Subject 3				
Subject 4				
Subject 5				

Fig. 9.9 Comparison of automatic and manual segmentation of LCPR for five different subjects

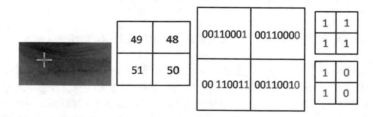

Fig. 9.10 Illustration of bit-plane slicing procedure

the image constitutes a particular bit-plane. The process of bit-plane slicing an 8-bit image is illustrated in Fig. 9.10.

Figure 9.11 shows series of eight bit-plane slices derived over input gray-scaled LCPR images for five different subjects. Higher order bits contain visually significant information of LCPR while lower order bits comprise more subtle details. A subset of eight bit-planes is sufficient to represent the image without significant loss of useful information. Bit-planes 1–4 correspond to fine and minute textural details, and bit-planes 7–8 consist of coarse information representing overall pictorial details. Bit-planes 5 and 6 contain more discernible information imperative to describe LCPR region. Useful LCPR features including details of skin texture and structure are more prominent in bit-plane 5 in comparison with bit-plane 6 visually. The following section justifies bit-plane 5 selection technically in two realms, namely: structural similarity of bit-plane image with grayscale image and density of texture content in each bit-plane required for periocular authentication.

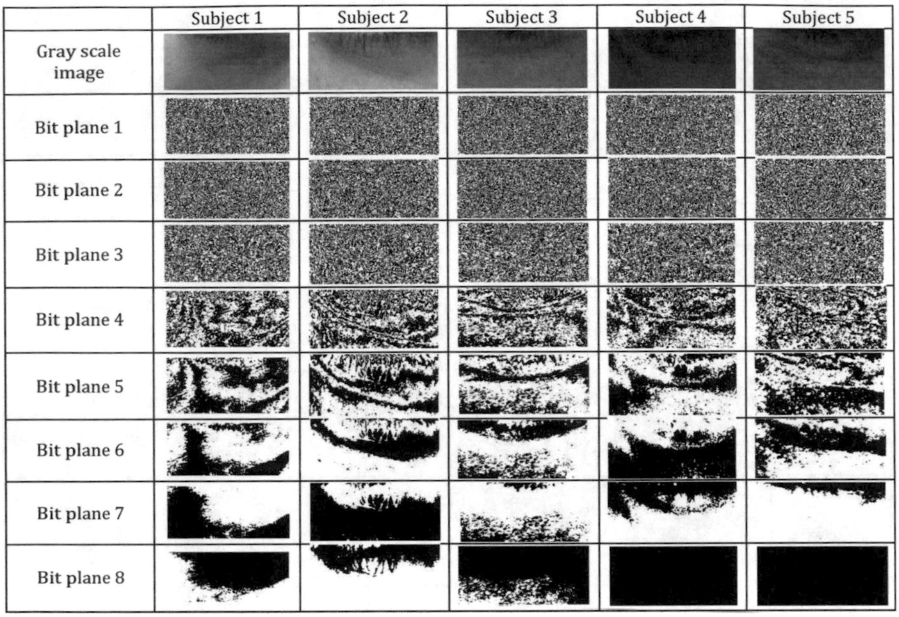

	Subject 1	Subject 2	Subject 3	Subject 4	Subject 5
Gray scale image					
Bit plane 1					
Bit plane 2					
Bit plane 3					
Bit plane 4					
Bit plane 5					
Bit plane 6					
Bit plane 7					
Bit plane 8					

Fig. 9.11 Eight grayscale bit-planes obtained for five different subjects

Bit-Plane 5 Selection Justification Using Structural Similarity Index

Visual quality of the eight bit-planes is quantified by assessing quality of bit-plane images in comparison to the corresponding grayscale LCPR image. A possible approximation of visual quality is the measure of change in structural information between the two images. Wang et al. described structural similarity index (SSIM) as an efficient measure of visual quality assessment [37]. Among the state-of-the-art image quality assessment index, SSIM has been proved to be better objective quality assessment metric [11]. The method performs the computation by independently comparing three structure components. The illumination component of bit-plane image I_1 is compared with grayscale image I_2 by using Eq. 9.2.

$$l(I_1, I_2) = \frac{2\mu_{I_1}\mu_{I_2} + C_1}{\mu_{I_1}^2 + \mu_{I_2}^2 + C_1} \tag{9.2}$$

μ_{I_1} and μ_{I_2} represent the mean intensity values of the two images. C_1 is a constant included to avoid instability resulting when $\mu_{I_1}^2 + \mu_{I_2}^2$ approximates to zero. Contrast comparison is expressed in terms of standard deviation values of the two images, σ_{I_1} and σ_{I_2}. The corresponding function is represented by Eq. (9.3). C_2 is a constant similar to C_1.

$$c(I_1, I_2) = \frac{2\sigma_{I_1}\sigma_{I_2} + C_2}{\sigma_{I_1}^2 + \sigma_{I_2}^2 + C_2} \tag{9.3}$$

Structural information being independent of luminance, structure component is computed by subtracting luminance and normalizing contrast information. Structure comparison function is given by Eq. (9.4) with a constant C_3 similar to C_1 and C_2.

$$s(I_1, I_2) = \frac{\sigma_{I_1 I_2} + C_3}{\sigma_{I_1}\sigma_{I_2} + C_3} \tag{9.4}$$

SSIM is obtained by combining the three comparison functions and approximating $C_3 = \frac{C_2}{2}$ for simplicity. Equation 9.5 represents the SSIM measure. Image statistics and SSIM values are computed locally within a window. Local SSIM values obtained by moving the local window over the entire image is averaged to compute the SSIM index.

$$\text{SSIM} = \frac{(2\mu_{I_1}\mu_{I_2} + C_1)(2\sigma_{I_1 I_2} + C_2)}{(\mu_{I_1}^2 + \mu_{I_2}^2 + C_1)(\sigma_{I_1}^2 + \sigma_{I_2}^2 + C_2)} \tag{9.5}$$

SSIM is implemented on three sets of data, namely: benchmark UBIRISv2 dataset, BMSCE 2D periocular images retrieved using high resolution; Canon EOS Mark III camera; and low resolution periocular images captured from MacBook Pro laptop. Figure 9.12 shows plot of SSIM values computed for 150 subjects from UBIRISv2 datasets over different bit-planes using grayscale LCPR image as reference. The values for bit-plane 5 is larger in Fig. 9.12a, with the peaks raising higher than other bit-plane values. The large values indicate that bit-plane 5 image is more structurally competent to represent original LCPR image. Smaller magnitudes of SSIM values correspond to lower order bit-planes indicating the presence of acute information regarding the overall LCPR structure. Figure 9.12b shows that bit-planes 6, 7, and 8 have high SSIM. The values indicate that overall structure of LCPR in the higher order bit-planes is similar to the original image. However, minute details of the texture corresponding to structural complexity of edges and ridges in LCPR are not represented. Texture is an important attribute for matching LCPR images in addition to structural component. Density of texture in each bit-plane image is used as an additional measure to justify contribution of bit-plane 5 for periocular authentication.

Bit-Plane 5 Selection Justification Using Texture Information

The density of texture present in bit-planes of LCPR is determined by computing fractal dimensions. Fractal dimension is regarded as a useful index to characterize texture complexity and details of geometric forms. Complex surfaces are associated with rough texture, and are represented with higher values of fractal dimensions. Smoother surfaces with finer texture have lower fractal dimensions. The "box

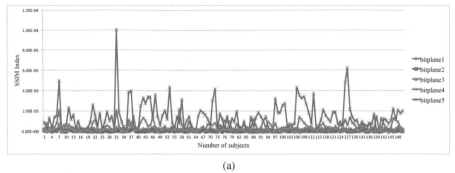

(a)

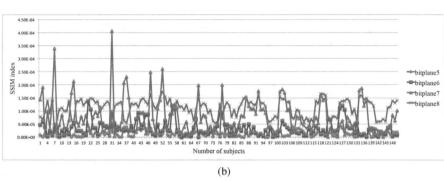

(b)

Fig. 9.12 SSIM values computed for (**a**) bit-planes1–5 and (**b**) bit-planes 5–8 for 150 subjects

counting approach" is implemented on the eight LCPR bit-plane images to estimate the fractal dimensions [22]. The method is based on Mandelbrot's self-similarity concept. A pattern is defined to be self-similar if the pattern is a union of N number of distinct, non-overlapping copies of itself. Each copy is similar to the original and scaled down by a ratio r in each dimension. Fractal dimension, D, of an image is expressed using Eq. (9.6) in box counting approach.

$$D = \frac{\ln N}{\ln \frac{1}{r}} \tag{9.6}$$

Figure 9.13 shows plots of fractal dimension computed over different bit-planes of LCPR images corresponding to 150 subjects of UBIRISv2 dataset. Higher values of fractal dimensions are observed in Fig. 9.13a for lower order bit-planes indicating higher density of texture information in bit-planes 1–4. Comparing bit-plane 5 with bit-planes 6 and 7 in Fig. 9.13b, bit-plane 5 incorporates more textural information. Abstraction of minute LCPR details is least in bit-plane 8 as depicted in Fig. 9.13c.

In accordance with above discussions, bit-planes 1–4 do not incorporate sufficient structural information and bit-planes 6–8 do not incorporate sufficient LCPR textural content. A combination of overall LCPR structure coupled with fine textural

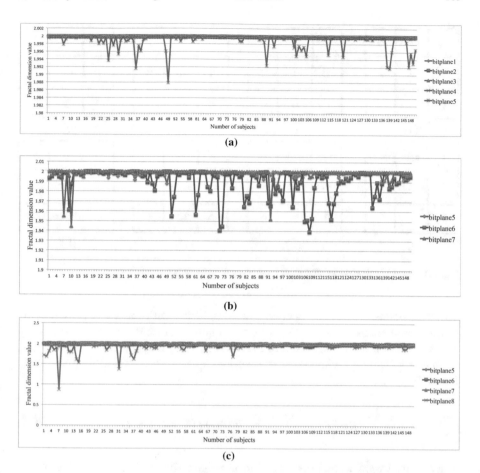

Fig. 9.13 Plot of (**a**) fractal dimensions for bit-planes 1-5, (**b**) fractal dimensions for bit-planes 5-7, and (**c**) fractal dimensions for bit-planes 5–8

details is adequately represented by bit-plane 5. Numerical values of SSIM and fractal dimensions for bit-plane 5 commensurate with the theoretical discussion and are selected as the dominant bit-plane for subsequent LCPR matching.

Dominant Bit-Plane LBP Feature Extraction Using Radial Filters

Feature extraction on the dominant bit-plane is implemented using a filter-based approach. A radial filter bank is constructed as a set of oriented derivative filters with the thresholded output as equivalent to local binary operator. The filter coefficients computed are equivalent to weights of binary interpolation of pixel values at

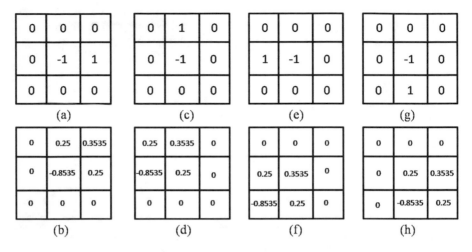

Fig. 9.14 **(a–h)** LBP-based radial filters corresponding to eight neighborhoods

sampling points of an LBP operator. In compliance with the work presented by
Matti et al., kernels shown in Fig. 9.14 are employed for LBP implementation [3].
The kernels represent LBP operator in a circular $(8,1)$ neighborhood. Convolution
of an $m \times n$ 2D image with the radial filter w_i for $i = 1, 2 \ldots .8$ is represented using
Eq. (9.7).

$$g(x, y) = \sum_{s=-m}^{m} \sum_{t=-n}^{n} w(s, t) f(x + s, y + t) \qquad (9.7)$$

x and y are varied so that each pixel in w neighborhood visits every pixel in the
image f. At each pixel co-ordinate (x, y), convolution stage involves computing the
product of the value f at that point and the corresponding radial filter component
$w(s, t)$. Multiplication is followed by the addition of neighborhood products to a
single value that replaces the center pixel (x, y). The process is repeated over the
entire image. The response of the filter at location (x, y) gives the signed differences
of the center pixel and the sampling point corresponding to the filter.

Convolution of radial filter bank on the dominant bit-slice benefits reduced
computational complexity. On slicing the grayscale image f into individual bit-
planes b_n binary images are resulted, where n is the number of bits used to represent
the image. The fact that bit-plane image constitutes pixel values of only $1's$ and
$0's$, Eq. (9.7) reduces to Eq. (9.8). Subsequent convolution operation involves only
sequences of addition operations.

$$g(x, y) = \sum_{s=-m}^{m} \sum_{t=-n}^{n} w(s, t) \quad \forall \quad b(x + s, y + t) = 1 \qquad (9.8)$$

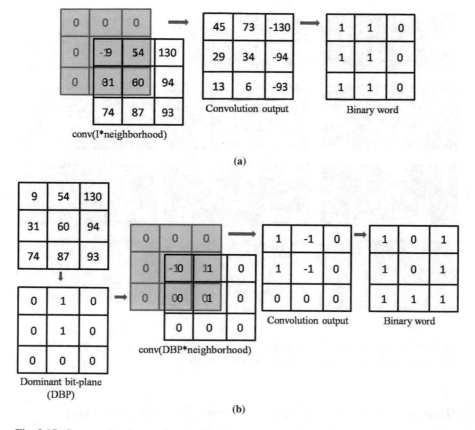

Fig. 9.15 Computational stages involved in (**a**) traditional LBP and (**b**) dominant bit-plane LBP

In reference to sample computation shown in Fig. 9.15, it is observed that traditional LBP mechanism requires a maximum of $(4 - l - 1)$ number of addition operations and $(4 - l)$ number of multiplication operations for each convolution operation. l represents the number of "zero" elements in the input submatrix at positions corresponding to non-zero elements of the kernel matrix. In contrast, convolution using dominant bit-plane largely reduces mathematical operations to only $(4 - l - 1)$ number of additions and no multiplication operations.

Figure 9.16 shows dominant bit-planes and the corresponding dominant bit-plane LBP texture maps for four different subjects. Dominant bit-plane being able to encapsulate predominant periocular texture, local primitives including curved edges, spots, and flat areas of periocular region are efficiently encoded into dominant bit-plane LBP features.

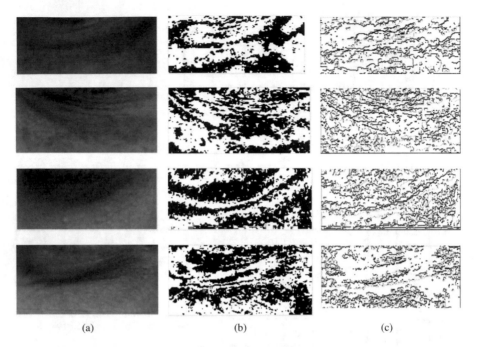

Fig. 9.16 (a) LCPR regions for four different subjects, (b) corresponding dominant bit-planes, and (c) corresponding dominant bit-plane LBP texture maps

Determination of Dominant Bit-Plane LBP Feature Vectors

Dominant bit-plane LBP texture features, computed according to the discussions of Sect. 9, are used to construct 256-bin histograms. Figure 9.17 shows dominant bit-plane LBP histograms generated for two samples of four different subjects. Graphs indicate that certain bins exhibit extreme values and lead to difficulty in differentiating the feature vectors among the subjects. Numerical values of the histograms are shown in Fig. 9.18, to clearly demonstrate the presence of intermittent values. The figure shows that certain bins hold lower extreme values corresponding to a range of $(0, 10)$ and certain other bins exhibit exorbitantly high extremes, raised to multiples of 10^3 or 10^4. The extreme values appear as outliers during matching and largely hinder comparison of histograms across different subjects. Imperative course of action to facilitate reliable matching is to

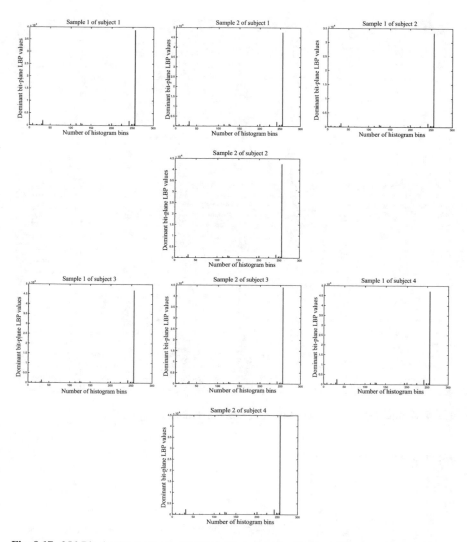

Fig. 9.17 256-Bin dominant bit-plane LBP histograms derived for four different subjects with two samples each

eliminate outliers. Post-processing, using outlier removal procedure, is performed over dominant bit-plane LBP histograms to accommodate moderate values useful and sufficient for subsequent classification. Details of the process are described in Algorithm 4.4.1. Figure 9.19 illustrates the three stages of outlier removal procedure pictorially.

P4_1	P4_2	P5_1	P5_2	P6_1	P6_2	P7_1	P7_2
223	217	141	129	122	157	114	130
278	228	133	133	155	174	191	245
2	3	2	0	1	1	0	1
6	6	2	0	4	4	4	7
136	127	109	109	87	103	121	134
49	38	28	31	20	27	38	35
5	6	2	0	6	0	4	9
446	325	274	234	268	321	409	401
1	2	2	0	0	0	3	2
2	1	1	1	4	1	0	3
0	0	0	1	0	0	0	0
1	0	0	0	0	0	0	0
2	1	1	4	3	2	3	7
1120	849	767	667	626	779	1078	1197
0	0	1	4	1	2	0	4
74	65	50	52	27	44	57	62
6	4	5	6	1	1	2	4
40	31	33	29	25	26	34	30
0	0	0	0	0	0	0	1
357	300	319	249	233	291	358	327
13	6	4	4	19	7	25	29
52	54	49	52	31	40	48	52
4	2	1	0	0	0	0	0
19	12	16	10	11	15	18	13
47	51	36	30	24	40	61	31
416	333	291	222	239	294	325	328
21	21	16	17	12	17	18	31
10832	9067	10335	7809	11411	11712	10476	11087

Fig. 9.18 Raw dominant bit-plane LBP values for four different subjects

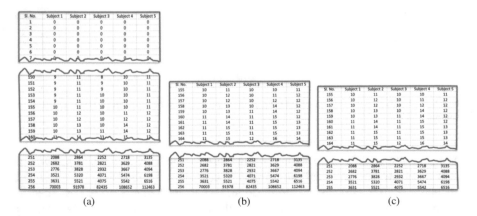

(a)   (b)   (c)

Fig. 9.19 Illustration of steps involved in outlier removal process. (**a**) Sorting of histogram bins. (**b**) Elimination of lower extreme values of histogram. (**c**) Elimination of higher extreme values of histogram

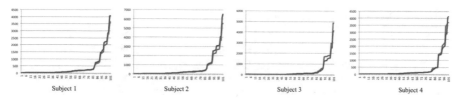

Fig. 9.20 Post-processed dominant bit-plane LBP feature vectors for two samples of four different subjects

---

4.4.1 Algorithm for outlier removal process

For N number of subjects, with LCPR image $\{f\}$ and bit-plane image $\{f_{bp}\}$,

- $\{f\} = \cup\{f_{bp_r}\}, r = 1, 2, 3, 4, 5, 6, 7, 8$
- Dominant bit-plane, $\{f_{Dbp}\}$ is determined as

$$\{f_{Dbp}\} = \{f\} - \{f_{bp_r} < SSIM_{thresh} \| f_{bp_r} < D_{thresh}\}$$

- dominant bit-plane LBP image g, using radial filter w, is obtained as
  - $g_{m \times n} = w * f_{m \times n}$
- dominant bit-plane LBP histogram, LBP_{DB} is computed as,

$$LBP_{DB} = \{v_i\}_{i=1:256} \ni \sum_{t=1}^{256} g_t = m \times n$$

- For i number of bins in each histogram,
  - Histograms are arranged as $LBP_{DB_{sorted}} = \{v_1 < v_2 < \cdots < v_{256}\}$
  - Lower extreme values are removed as

$$LBP_{DB_{Ltruncate}} = \subset \{LBP_{DB_{sorted}}\} = \{v_i\}|_{v_i < 10} \approx \subset \{v_i\}_{i > 155}$$

  - Higher extreme values are removed as

$$LBP_{DB_{Htruncate}} = \subset LBP_{DB_{Ltruncate}} = \{v_i\}|_{v_i > 10^4} \approx \subset \{v_i\}_{i > 255}$$

- Post processed LBP feature vector, $LBPDB_{fv} = LBP_{DB_{Htruncate}}$
$$\approx \subset \{v_i\}_{i=1:255}$$

Post-processed feature vectors for four different subjects using two samples each are shown in Fig. 9.20. Resulting 100-element feature vector (using values from 156th to 255th bin) is presented to the classifier. During classification, a probe image is matched with a set of target images using a similarity measure formulated by city-block distance.

Experiments

Experimental Datasets

The framework proposed in the current work is tested over periocular and LCPR images retrieved from benchmark UBIRISv2 database along with high resolution images acquired from Canon camera and low resolution images acquired using

Table 9.1 Details of image acquisition setups

Database	Image resolution	Description of periocular images	Description of unconstrained environment
UBIRISv2 benchmark database	Horizontal: 72 dpi, Vertical: 72dpi	Non-uniform sizes of periocular regions for each subject	Variations in distance, scale, and occlusion
BMSCE 2D periocular images	5760 × 3840	Uniform sizes of periocular region for all subjects	Variations in expression, pose, and occlusion
Low resolution periocular images	1080 × 720	Uniform sizes of periocular regions for all subjects	Variations in expression: smile, surprise, and anger, occlusions due to flapped eyes and hair

MacBook Pro laptop camera. Canon EOS 5D Mark III camera is used to acquire high resolution periocular images and iSight camera of MacBook Pro laptop is used to obtain low resolution periocular images. Table 9.1 presents image descriptions of the three datasets.

UBIRISv2 Dataset

UBIRISv2 database constitutes visible wavelength eye images of 261 subjects, acquired on-the-move, at-a-distance, using Canon EOS 5D camera. The database poses several challenges in scale, occlusion, and illumination [30]. Fifteen images are available for each subject with only a subset of images having representable periocular region and others mostly display iris region. An experimental dataset for each subject is constructed using six images with sufficient periocular region and no spectacles. LCPR is segmented from each periocular image of the dataset by employing active contour segmentation approach discussed in Sect. 9. Experiments over UBIRISv2 database demonstrate that up to 99.5% of automatically segmented ROIs are similar to the manually segmented regions.

High Resolution Images

High resolution samples of BMSCE 2D periocular images were acquired using Canon EOS 5D Mark III camera. The details of image acquisition setup are tabulated in Table 9.2. Even though UBIRISv2 database was constructed using a similar camera, major purpose of UBIRISv2 database was facilitation of iris recognition and the area of periocular region available for each image varied largely. Interest of the current database is to avail complete and uniform area of periocular region for all images. Current acquisition framework captured facial images at resolution 5760 × 3840. A set of 1620 high resolution images were

Table 9.2 Image acquisition setup for BMSCE 2D periocular database

Setting of camera parameters	Maximum resolution—22.3 Megapixels (5760 × 3840, AUTO FOCUS, ISO AUTO, and JPEG format
Details of acquisition environment	The imaging framework was installed in a room under both artificial and natural lighting conditions. The subjects were imaged at three distances: 1 m, 1.5 m, and 2 m. Markings were made on the walls at angles $+30^0$ and -30^0 from a reference 0^0 line and the subjects were asked to look at the markings for image acquisition
Particulars of manual cropped periocular images	width = 600 pixels, height = 500 pixels, format = tiff

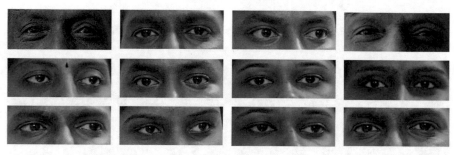

Fig. 9.21 Samples of BMSCE 2D periocular images

collected from 45 different subjects in two different sessions. 36 images exhibiting varied distances, expressions, and pose are constructed in two different sessions for each subject. Occlusions included flapped eyes and the presence of hair. Sample periocular images constituting the experimental dataset are shown in Fig. 9.21. Left and right periocular regions were manually cropped along the dashed lines as shown in Fig. 9.22. The regions were bounded by eyebrows above and start of cheekbone below. LCPR images are segmented from each periocular region using the automatic segmentation algorithm discussed in Sect. 9. 99.7% of automatically segmented LCPRs matched with manually segmented images.

Low Resolution Images

Low resolution images acquired using built-in iSight camera, Apple Inc of Mac-Book Pro laptop. The webcam captures facial images at 1080 × 720 resolution. Varied expressions and occlusions are considered during imaging. Periocular images constituting both eyes are manually cropped for each subject. Left and right periocular regions are subsequently extracted and stored. A total of 120 periocular images are collected from twenty subjects. A set of six images comprising 2 neutral expression, 1 with eyes flapped and 3 with expression variation (smile, surprise, and anger) are acquired for each subject. Sample images constituting the dataset are shown in Fig. 9.23.

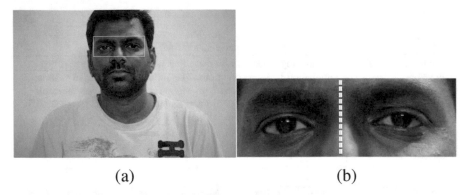

(a)   (b)

Fig. 9.22 Segmentation of periocular image from the facial image. (**a**) Original facial image. (**b**) Periocular image

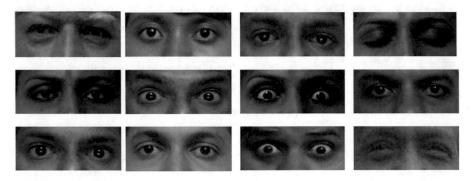

Fig. 9.23 Samples of low resolution periocular images acquired using laptop built-in camera

Results and Discussion

Experimental Validation of Dominant Bit-Plane

An experimental validation is conducted to reemphasize justification of the fifth bit-plane as the dominant bit-plane. LCPR samples of fifteen different subjects from UBIRISv2 dataset are employed for testing. Table 9.3 tabulates authentication results achieved over different bit-planes of LCPR, using the dominant bit-plane LBP framework. The table shows that FAR and FRR values computed over the eight bit-planes are in compliant with the discussion presented in Sect. 9. The system performance is best achieved by using bit-plane 5 over others. Higher order bit-planes witness reduced accuracy due to loss of finer details and abstraction of only global pictorial information. Utilizing lower order bit-planes affects the performance due to poor representation of discernible features among different subjects, despite encompassing fine and minute texture details. Bit-plane 5 is more promising with acceptable performance emphasizing the ability to represent useful and discernible periocular information most appropriately.

Table 9.3 FAR and FRR computed for different bit-planes of LCPR

Bit-plane	FRR	FAR	Bit-plane	FRR	FAR
Bit-plane 1	2.58%	2.08%	Bit-plane 5	1.00%	1.00%
Bit-plane 2	2.50%	2.50%	Bit-plane 6	0.90%	5.50%
Bit-plane 3	0.90%	5.33%	Bit-plane 7	2.50%	4.50%
Bit-plane 4	0.90%	4.83%	Bit-plane 8	4.16%	4.33%

Authentication Accuracies

UBIRISv2 Database Figure 9.24a shows ROC plots obtained for left LCPR, right LCPR, left periocular and right periocular regions. The curves show that an authentication accuracy of 98.2% and 98.3% is achieved, respectively, for left and right LCPR images. Using entire periocular region, an accuracy of 98.7% and 99% is attained for left and right regions, respectively. The results indicate that no significant degradation in performance is witnessed with reducing the region of interest to only a portion of periocular region, LCPR.

High Resolution Images Figure 9.24b shows ROC curves obtained for LCPR and entire periocular images using DB-LBP methodology. An authentication accuracy of up to 99.5% is achieved by employing LCPR images. Results tabulated in Table 9.4 show that LCPR regions outperform periocular region particularly under varied expressions. An improvement of 2.9% is witnessed by LCPR images in the presence of occlusions. Increased performance of LCPR is accounted for the reduced deformation during change in expression, eyeball movement, and the presence of occlusion. Authentication attained under varied illumination and pose is in equivalence with the results obtained using entire periocular region.

Low Resolution Images Figure 9.24c depicts ROC curves obtained for LCPR and corresponding periocular images. An authentication accuracy of 91% and 91.2% is attained for left and right LCPR images. Left and right periocular images achieve 90.1% and 91%, respectively. Results show that authentication using the suggested sub-region of periocular area is more efficient than using the entire region around the eye. The characteristic of LCPR, being the least deformed portion of periocular region, enhances the performance of the proposed authentication system.

Table 9.5 discusses performance of the state-of-the-art periocular biometric systems using LBP and its variant feature extraction schemes. The schemes are evaluated under varied unconstrained acquisition environments. The table indicates that DB-LBP framework attains similar performance as the state-of-the-art periocular biometric systems with the benefits of reduced template size.

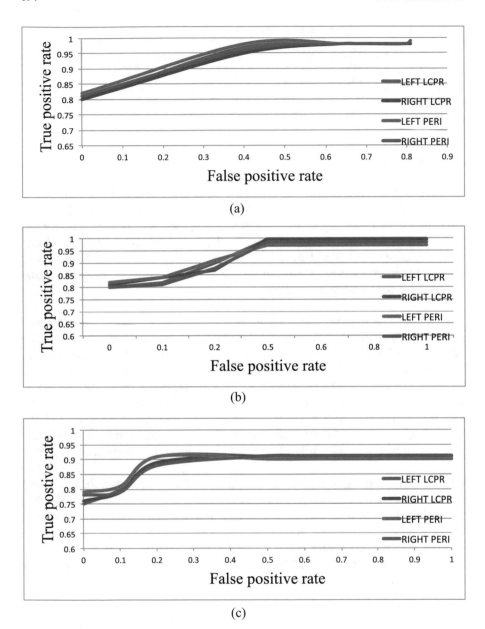

Fig. 9.24 ROC curves obtained for UBIRISv2, high resolution image and low resolution image datasets, using entire periocular region and using LCPR

Table 9.4 Results obtained using BMSCE 2D images under varied expressions, illumination, pose, and occlusions due to hair and flapped eyes

Variations in	Left LCPR	Right LCPR	Left peri	Right peri
Expressions	99.2%	99.5%	96.5%	97%
Illumination	97%	97.1%	97%	97.2%
Pose	97.1%	97.3%	97%	98.2%
Occlusion due to hair and flapped eyes	99.2%	99.4%	96%	96.5%

Table 9.5 Comparison of the state-of-the-art periocular authentication/recognition performances

Author	Methodology	Variability factors	Accuracy %
Lyle et al. [24]	LBP + SVM	Soft biometric classification	93.0
Adams et al. [1]	LBP + GEFE	Distance, lighting, expression	92.2
Samarth et al. [14]	GIST + CLBP	Distance	78.6
Park et al.[29]	GO + LBP + SIFT	Pose, occlusion, age	80.8
Xu et al. [19]	WLBP + KCFA	Light, pose, expression	60.7
Joshi et al. [4]	Gabor + LBP	Light, distance, pose	89.7
Xu et al. [20]	DT − LBP	Expression, occlusion, distance	75.1
Bakshi et al. [13]	Walsh Hadamard LBP	Light, pose, distance	65.76
Uzzair et al. [35]	LBP+PCA	Light, distance, pose, spectrum	99.8

Comparison of Entire Face, Periocular Region, and LCPR Using DB-LBP

Face images available during construction of high and low resolution 2D periocular dataset are experimented to compare performance of face with LCPR. Sample face images used for testing are shown in Fig. 9.25. Corresponding authentication accuracies are tabulated in Table 9.6. Results show that using entire facial information does not improve the accuracy better than using only LCPR area. Deformation of lips and cheeks in the lower portion of face during expression variation affects face matching results. Loss of facial information due to left and right orientations of head in pose variant images presents an additional constraint on improving the performance. The results enunciate that supplementing LCPR with other portions of periocular region does not upsurge the performance largely. Instead, an increased classification rate is achieved for LCPR images of BMSCE 2D periocular dataset, indicating its robustness to varied expression, pose, and the presence of hair occlusions. Regardless of low resolution, expression variant images acquired using laptop camera have been authenticated successfully and have shown an improvement over periocular results.

Table 9.7 tabulates dominant bit-plane LBP computation time for different regions of interests. Computation time is determined by running Matlab implementation on a MacBook Pro laptop machine with 2.5 GHz Intel Core i5 processor and 4GB RAM. Matlab R2014a is used as programming environment. The table

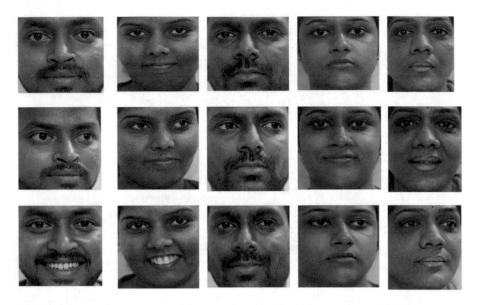

Fig. 9.25 Sample face images used for testing impact of facial information with LCPRs

Table 9.6 Comparison of DB-LPB authentication results using LCPR, periocular region, and entire face images

Dataset	LCPR	Periocular region	Entire face
UBIRISv2	98.3%	99%	NA
BMSCE 2D periocular images	99.5%	98.2%	96%
2D periocular image	91.2%	91%	89.0%

Table 9.7 Time required for processing different regions of interest using DB-LBP

Region of interest	Time elapsed
Face images	0.7321 s
Periocular images	0.2737 s
LCPR images	0.1825 s

Table 9.8 Comparison of computation times of DB-LBP with traditional LBP

Method	Image size	Time elapsed (s)
Dominant bit-plane LBP	150×600	0.1530
Filter-based LBP	150×600	0.2356
Pixel-wise LBP	150×600	19.3002

indicates that time required for processing LCPR images is approximately $\frac{1}{4}th$ of the time required for processing LCPR images. Computation time of periocular region is 1.5 times the time required for LCPR images.

Conclusion

Two prime contributions are made in the present work towards implementation of authentication systems, specifically for memory restricted applications: (a) proposition of dominant bit-plane LBP as a computationally efficient variation of conventional LBP method and (b) investigation of reduced template, LCPR, to authenticate an individual in an unconstrained acquisition environment. The study shows that certain bit-plane images efficiently encapsulate perceptual textural characteristics useful for classification. The idea of extracting LBP features on a dominant bit-plane is interesting as the required operations for computing the feature vector are simplified suiting needs of smart device applications. Table 9.8 tabulates times elapsed for LCPR feature extraction techniques using conventional LBP and the proposed dominant bit-plane LBP. The table indicates that the proposed scheme accomplishes feature extraction with a least time of only 0.153 s for an input image of size 150 × 600. An effective run time ratio of 1.54 is achieved between dominant bit-plane LBP and filter-based LBP approaches. Further, employing only potential portions of periocular area adds to the benefit of reduced computational cost and space, facilitating storage of large databases. Resistance to pose variation and expression variation is attained without any pre-processing or pose correction algorithms prior to feature extraction.

References

1. Adams, J., Woodard, D. L., Dozier, G., Miller, P., Bryant, K., & Glenn, G. (2010). Genetic-based type II feature extraction for periocular biometric recognition: Less is more. In *20th International Conference on Pattern Recognition*. Piscataway: IEEE.
2. Ahonen, T., Hadid, A., & Pietikäinen, M. (2006). Face description with local binary patterns: Application to face recognition. *IEEE Transactions on Pattern Analysis and Machine Intelligence, 28*(12), 2037–2041.
3. Ahonen, T., & Pietikäinen, M. (2008). A framework for analyzing texture descriptors. In *VISAPP 2008: Proceedings of the Third International Conference on Computer Vision Theory and Applications* (Vol. 1, pp. 507–512). Setubal: Institute for Systems and Technologies of Information, Control and Communication.
4. Akanksha, J., Abhishek, G., Renu, S., Ashutosh, S., & Zia., S. (2014). Periocular recognition based on Gabor and Parzen PNN. In *2014 IEEE International Conference on Image Processing (ICIP)* (pp. 4977–4981). Piscataway: IEEE.
5. Alabady, S. A., & Al-Turjman, F. (2018). Low complexity parity check code for futuristic wireless networks applications. *IEEE Access Journal, 6*(1), 18398–18407.
6. Alabady, S. A., & Al-Turjman, F. (2018). A novel security model for cooperative virtual networks in the IoT era. *International Journal of Parallel Programming*, 1–16.
7. Al-Turjman, F., & Alturjman, S. (2018). Confidential smart-sensing framework in the IoT era. *The Journal of Supercomputing, 74*(10), 5187–5198.
8. Al-Turjman, F., & Alturjman, S. (2018). Context-sensitive access in industrial internet of things (IIoT) healthcare applications. *IEEE Transactions on Industrial Informatics, 14*(6), 2736–2744.

 9. Al-Turjman, F., Hasan, M. Z., & Al-Rizzo, H. (2018). Task scheduling in cloud-based survivability applications using swarm optimization in IoT. *Transactions on Emerging Telecommunications Technologies*, e3539.
10. Ambika, D. R., Radhika, K. R., & Seshachalam, D. (2016). Periocular authentication based on fem using Laplace–Beltrami eigenvalues. *Pattern Recognition Journal, 50*(C), 178–194.
11. Bae, S. H., & Kim, M. (2015). A novel SSIM index for image quality assessment using a new luminance adaptation effect model in pixel intensity domain. In *Visual communications and image processing (VCIP)*. Piscataway: IEEE.
12. Bakshi, S., Sa, P. K., & Majhi, B. (2013). Optimized periocular template selection for human recognition. *BioMed Research International, 2013*, 14. https://doi.org/10.1155/2013/481431.
13. Bakshi, S., Sa, P. K., & Majhi, B. (2015). A novel phase-intensive local pattern for periocular recognition under visible spectrum. *Biocybernetics and Biomedical Engineering, 35*(1), 30–44.
14. Bharadwaj, S., Bhatt, H., Vatsa, M., & Singh, R. (2010). Periocular biometrics: When iris recognition fails. In *Fourth IEEE International Conference on Biometrics Compendium* (pp. 1–6). Piscataway: IEEE.
15. Boddeti, V. N., Smereka, J. M., & Kumar, B. V. (2011). A comparative evaluation of iris and ocular recognition methods on challenging ocular images. In *2011 International Joint Conference on Biometrics (IJCB)* (pp. 1–8). Piscataway: IEEE
16. Chan, T., & Vese, L. (2001). Active contour without edges. *IEEE Transactions on Image Processing, 10*, 266–277.
17. Hollingsworth, K., Bowyer, K. W., & Flynn, P. J. (2010). Identifying useful features for recognition in near-infrared periocular images. In *Fourth IEEE International Conference on Biometrics: Theory Applications and Systems (BTAS)* (pp. 1–8). Piscataway: IEEE
18. Hollingsworth, K., Darnell, S. S., Miller, P. E., Woodard, D. L., Bowyer, K. W., & Flynn, P. J. (2012). Human and machine performance on periocular biometrics under near-infrared light and visible light. *IEEE Transactions on Information Forensics and Security, 7*(2), 588–601.
19. Juefei, X., & Savvides, M. (2012). Unconstrained periocular biometric acquisition and recognition using COTS PTZ camera for uncooperative and non-cooperative subjects. In *IEEE Workshop on Applications of Computer Vision, WACV* (pp. 201–208). Piscataway: IEEE.
20. Juefei, X., & Savvides, M. (2014). Subspace based discrete transform encoded local binary patterns representations for robust periocular matching on NIST's face recognition grand challenge. *IEEE Transactions on Image Processing, 23*, 3490–3505. https://doi.org/10.1109/TIP.2014.2329460.
21. Kanumuri, T., Dewal, M. L., & Anand, R. S. (2014). Progressive medical image coding using binary wavelet transforms. *Signal, Image and Video Processing, 8*(5), 883–899.
22. Li, J., Du, Q., & Sun, C. (2009). An improved box-counting method for image fractal dimension estimation. *Pattern Recognition Journal, 42*(11), 2460–2469.
23. López, M. B., Nieto, A., Boutellier, J., Hannuksela, J., & Silvén, O. (2014). Evaluation of real-time LBP computing in multiple architectures. *Journal of Real-Time Image Processing*, 1–22.
24. Lyle, J. R., Miller, P. E., Pundlik, S. J., & Woodard, D. L. (2010). Soft biometric classification using periocular region features. In *Fourth IEEE International Conference on Biometrics Compendium* (pp. 1–7). Piscataway: IEEE.
25. Miller, P. E., Lyle, J. R., Pundlik, S. J., & Woodard, D. L. (2010). Performance evaluation of local appearance based periocular recognition. In *Fourth IEEE International Conference on Biometrics Compendium* (pp. 1–6). Piscataway: IEEE.
26. Ojala, T., Pietikainen, M., & Harwood, D. (1994). Performance evaluation of texture measures with classification based on Kullback discrimination of distributions. In *International conference on Pattern Recognition (ICPR)* (Vol. 1, pp. 582–585). Piscataway: IEEE.
27. Padole, N. C., & Proença, H. (2012). Periocular recognition: Analysis of performance degradation factors. In: *5th IAPR International Conference on Biometrics (ICB)* (pp. 439–445). Piscataway: IEEE. https://doi.org/10.1109/ICB.2012.6199790.
28. Park, U., Arun, R., & Jain, A. K. (2009). Periocular biometrics in the visible spectrum: A feasibility study. In *IEEE 3rd International Conference on Biometrics: Theory, Applications, and Systems, BTAS* (Vol. 6, pp. 28–30). Piscataway: IEEE.

29. Park, U., Ross, A., & Jain, A. K. (2011). Periocular biometrics in the visible spectrum. In *IEEE Transactions on Information Forensics and Security*. Piscataway: IEEE.
30. Proenca, H., Filipe, S., Santos, R., Oliveira, J., & Alexandre, L. A. (2010). The ubiris.v2: A database of visible wavelength iris images captured on-the-move and at-a-distance. *IEEE Transactions on Pattern Analysis and Machine Intelligence, 32*(8), 1529–1535.
31. Proença, H., Neves, J. C., & Santos, G. (2014). Segmenting the periocular region using a hierarchical graphical model fed by texture/shape information and geometrical constraints. In *IEEE International Joint Conference on Biometrics (IJCB)*. Piscataway: IEEE.
32. Rousson, M., & Paragios, N. (2002). Shape priors for level set representations. In *European Conference on Computer Vision* (pp. 78–92). Berlin: Springer.
33. Seok, O. B., Kangrok, O., & Ann, T. K. (2012). On projection-based methods for periocular identity verification. In *Industrial Electronics and Applications (ICIEA)* (pp. 871–876). Piscataway: IEEE.
34. Sharma, A., Verma, S., Vatsa, M., & Singh, R. (2014). On cross spectral periocular recognition. In *2014 IEEE International Conference on Image Processing* (pp. 5007–5011). Piscataway: IEEE.
35. Uzair, M., Mahmood, A., Mian, A., & McDonald, C. (2015). Periocular region-based person identification in the visible, infrared and hyperspectral imagery. *Neurocomputing, 149*, 854–867
36. Vazquez-Fernandez, E., Garcia-Pardo, H., Gonzalez-Jimenez, D., & Perez-Freire, L. (2011). Built-in face recognition for smart photo sharing in mobile devices. In *IEEE International Conference on In Multimedia and Expo (ICME)* (pp. 1–4). Piscataway: IEEE
37. Wang, Z., Bovik, A. C., Sheikh, H. R., & Simoncelli, E. P. (2004). Image quality assessment: From error visibility to structural similarity. *IEEE Transaction on Image Processing, 13*(4), 600–612.
38. Woodard, D. L., Pundlik, S. J., Lyle, J. R., & Miller, P. E. (2010). Periocular region appearance cues for biometric identification. In *2010 IEEE Computer Society Conference on Computer Vision and Pattern Recognition (CVPRW)* (pp. 162–169). Piscataway: IEEE.
39. Woodard, D., Pundlik, S., Miller, P., Jillela, R., & Ross, A. (2010). On the fusion of periocular and iris biometrics in non-ideal imagery. In *20th International Conference on Pattern Recognition*. Piscataway: IEEE.
40. Woodard, D., Pundlik, S., Miller, P., & Lyle, J. R. (2011). Appearance-based periocular features in the context of face and non-ideal iris recognition. In *Signal, Image and Video Processing* (Vol. 5). Berlin: Springer.

Chapter 10
Network Analysis of Dark Web Traffic Through the Geo-Location of South African IP Address Space

Craig Gokhale

Introduction

With the proliferation of cyberattacks and the increasing effectiveness of cybercriminals, organizations and indeed nation states are increasingly finding themselves vulnerable and at risk [1]. The coordinated attacks of ISIS, emergence of Ransomware, increasing online drug sales, and child pornography have led to questions being asked as to how these cybercrimes cannot be detected earlier. The answer lies in anonymous networks such as TOR. The Tor Project is a collection of software to allow anonymous communication and use of Internet technologies via a secured network utilizing application level encryption [2]. The technology stack makes it difficult for site owners and technology experts to identify and trace the client systems.

Previous studies [3–5] mostly involved reducing the complexity of this process by first reducing the set of relays or network routers to monitor, and then identifying the actual source of anonymous traffic among network connections that are routed via this reduced set of relays or network routers. A study by Burch [5] and Bauer [4] in this field reveals that there have been many more efforts to reduce the set of relays or routers to be searched than to explore methods for actually identifying an anonymous user amidst the network connections using these routers and relays. Few researchers have tried to comprehensively study a complete attack that involves reducing the set of relays and routers to monitor and identifying the source of an anonymous connection. Furthermore, to date, there is little or no evidence of any studies which characterize the usage of a real deployed anonymity service. Thus, this study presents observations and analyses obtained by participating in the TOR

C. Gokhale (✉)
University of Zululand, Teaching and Learning Centre, Durban, Kwa Zulul Natal, South Africa
e-mail: craiggokhale@gmail.com

© Springer Nature Switzerland AG 2020   201
F. Al-Turjman (ed.), *Smart Cities Performability, Cognition, & Security*,
EAI/Springer Innovations in Communication and Computing,
https://doi.org/10.1007/978-3-030-14718-1_10

network and identifying the illicit trade conducted by South Africans on the TOR network. More specifically, the purpose of the study is to investigate the use of TOR (and alternative anonymizer frameworks) within the South African context and specifically utilizing some of the weaknesses in these frameworks to identify the nature, source, and destination of South African traffic on the anonymous network.

Related Research

According to Murdoch [6] deanonymization of anonymous communication is a two-step process; the first step involves finding the anonymity set. In anonymity and privacy parlance, the anonymity set is the set consisting of anonymized entities (computers, humans, etc.), whose true identities have been hidden using some anonymization scheme. The observer of such a set observes the actual set and the set containing the former's anonymized identities but has no way to determine the relationship between the two sets. Murdoch [6] further notes that from the point of view of attacks against anonymous communication networks, involving identification of a certain anonymous client, every host or network on the Internet could be a potential victim. However, it is not feasible, for even powerful adversaries, to monitor each and every network or host, for identifying the source of anonymous traffic. Various research efforts have been made to reduce the set of hosts or network routers to monitor. In practice, some researchers have described how this anonymity set could be determined through a Sybil attack, wherein the adversary runs several malicious relays with the hope that some of these would be selected in users' connections and would aid an adversary to observe traffic entering and leaving the anonymization network. Other strategies, assuming powerful adversaries, involve observing traffic entering and leaving the network by observing traffic in vantage Autonomous Systems (ASes) or Internet Exchange Points (IXes), intervening the paths from various networks to the relays of anonymization networks [7–10].

The second part of the deanonymization process deals with finding the actual source of anonymous traffic by monitoring network connections that use the routers or network relays, that make up the anonymity set, and transport the victim traffic This generally involves some form of traffic analysis attack, wherein the adversary, having access to traffic in various networks, can correlate traffic transiting the anonymization relays with traffic flowing to (or from) the anonymity set, and identify the source of anonymous traffic. Low-latency anonymous communication systems geared towards semi-real-time applications, try to assure users' quality of service, by not modifying packet inter-arrival characteristics, such as delay and jitter. This makes them particularly vulnerable to traffic analysis attacks [11, 12].

Step 1: Finding the Anonymity Set

Researchers such as Pappas [13], Bauer [4], and Ovelier [14] in the past decade have explored various methods to determine the anonymity set. Some of these efforts included Sybil attacks, where the adversary runs malicious TOR entry and exit relays with the hope that they would get selected in circuits, and the node operators would be in a position to observe traffic entering and leaving the TOR network. Such an attack was first explored in 2007 [4], where the authors proposed attracting large fraction of TOR traffic by running malicious relays that advertised high available bandwidth. An anonymous client could select such malicious relays in entry and exit positions, which might be engaging in traffic analysis attack. Such attacks succeed because TOR relay selection is biased on advertised bandwidth [2].

In a related attack, Pappas [13] suggested creating looping circuits across non-malicious TOR relays and keeping relays busy, so as to prevent them from being selected in circuits. In the meanwhile, the adversary could run malicious relays that might advertise high bandwidth to increase their relative chance of being selected in circuits, especially while the benign nodes are busy serving malicious circuits specifically designed to keep them busy.

Ovelier [14] proposed the building of TOR paths, solely using nodes, known as guard nodes, so as to avoid such attacks. The threat of such Sybil attacks, involving malicious relays attracting users' traffic, is however only partially mitigated. Malicious relay operators could deploy nodes with high bandwidth for a certain period of time, so as to gain adequate trust, before launching in traffic analysis attacks.

Some researchers [7, 8], having studied the topology of the Internet, concluded that on average, 22% of TOR circuits originate from different subnets and that there are ASes which can observe traffic going towards a TOR entry node (from various subnets) and from exit nodes to some popular destinations, for example, popular search engines and free web-mail services. More recently [9], noted that a small number of compromised TOR relays that advertise high bandwidth and IXes observing both entry and exit traffic can deanonymize 80% of various types of TOR circuits within about 6 months.

In 2012, Murdoch [6] showed primarily through simulations, that a small set of IXes could observe traffic entering and leaving several TOR entry and exit nodes within the UK. He further revealed through simulations, that NetFlow, a traffic monitoring system installed in commodity routers, could be used to launch analysis attacks against TOR. The results were, however, mostly based on simulations, involving data obtained from by observing a single TOR relay.

The research reported above does not provide any intuition of the accuracy one can expect while practically using Net-Flow based statistics to deanonymize anonymous traffic. Thus, this research attempts to fill this gap by performing active traffic analysis to deanonymize TOR clients and explore the accuracy and practical issues involved in deanonymizing anonymous traffic.

Traffic analysis techniques have been explored in the past, for determining the anonymity set. For example, in 2012, Murdoch [10] developed the first practical traffic analysis attack against TOR and proposed a technique to determine the TOR relays involved in a circuit. The method involved a corrupt server, accessed by the victim client, and corrupt TOR client that could form one-hop circuits with arbitrary TOR nodes. The server modulates the data being sent back to the client, while the corrupt TOR node is used to measure delay between itself and TOR nodes. The inverse correlation between the perturbations in the client server traffic, deliberately introduced by the corrupt server, and the one-way delay, measured by the corrupt TOR client helped identify the relays involved in a particular circuit. Evans [15] demonstrated that the traffic analysis attack proposed by Murdoch [6] was no more applicable due to the large number of TOR relays, the large volume of TOR traffic, low end-to-end quality of service, and possible network bottleneck locations between the adversaries' vantage point and the victim relays. The above researcher proposed a method to amplify the network traffic by using circuits that repeatedly used the same relays and aided in easier identification of the relays.

Anonymous Network Communication Systems

Anonymous network communication systems enable users to hide their network identity (e.g., IP address) from their communication peers and also prevents network eavesdroppers to know the actual source or destination of messages. Most of these systems rely on sending traffic via one or more proxies, and may additionally encrypt traffic [16], to obfuscate the true source or destination of messages (as described ahead in detail). Such systems are often classified as low-latency and high-latency anonymous communication systems. Low-latency systems are designed to be efficient for semi-interactive applications such as web browsing and instant messaging. High-latency systems are geared towards delay tolerant applications such as e-mail. Low-latency network anonymization systems are further classified based on the routing paradigms they employ—those that are derived from Onion Routing [16] and those that are based upon Crowds [17]. Systems such as TOR [16], JAP [JAP,], and I2P [i2p,] employ deterministic routing, wherein the set of proxies through which the traffic is sent is known by the connection or session initiator (Fig. 10.1).

The client obtains a list of the available TOR relays from a directory service 1, establishes a circuit using multiple TOR nodes 2, and then starts forwarding its traffic through the newly created circuit [4], and One Swarm [19] employ probabilistic traffic routing schemes similar to Crowds. Each traffic forwarding relay in such a system randomly chooses to send the traffic either to the destination or to another relay in the system.

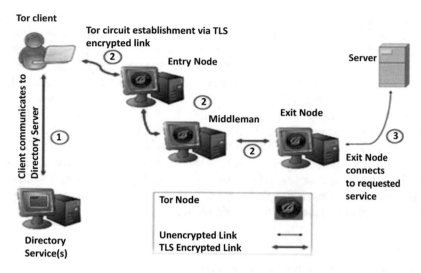

Fig. 10.1 Basics steps for communicating through TOR. Reproduced from: McCoy [18]

The Deep and Dark Web Defined

If we conceive the Web as a data ocean, most of us are interacting with the wavy, transparent, and easily navigable Surface Web. The Surface Web is the portion of the Web that has been crawled and indexed (and thus searchable) by standard search engines such as Google or Bing via a regular web browser. In the darkness below, beneath the electronic thermocline, are the abyssal depths of the Deep Web (also referred to as the Invisible Web or Hidden Web) the portion of the web that has not been crawled and indexed, and thus is beyond the sonar reach of standard search engines. It is technically impossible to estimate accurately the size of the Deep Web [1]. However, it is telling that Google currently the largest search engine has only indexed 4–16% of the Surface Web. The Deep Web is approximately 400–500 times more massive than the Surface Web. It is estimated that the data stored on just the 60 largest Deep Web sites alone are 40 times larger than the size of the entire Surface Web Growing rapidly within the Deep Web is the Dark Web (also referred to as the Dark Web, Dark Net, or Dark Internet) [20]. Originally, the Darkness referred to any or all network hosts that could not be reached by the Internet. However, once users of these network hosts started sharing files (often anonymously) over a distributed network that was not indexed by standard search engines, the Dark Web became a key part of the Deep Web. Unlike the traffic on the Surface Web or most parts of the Deep Web, most Dark Web sites can only be accessed anonymously. Preliminary studies have revealed that the Deep Web actually contains the largest expanding reservoir of fresh information on the Internet. These websites are usually narrower, but with much deeper content material, as compared to regular surface sites. Furthermore, because most of the materials are protected content, the overall quality of the content from the Deep Web is typically better and more valuable than that of the Surface Web. It is also estimated that more than 50% of the Deep Web content is located in topic-specific directories (www.thehiddenwiki.net), making it even more accessible and relevant to targeted searches. In addition the Deep Web and Dark Web are growing. Multiple technologies, such as ubiquitous computing, distributed/cloud computing, mobile computing, and sensor networks, have all contributed to the expansion of the Deep Web [21]. Advances in secure/anonymous web hosting services, crypto currency/Dark Wallet, and development of crimeware are further contributing to the growth of the Dark Web. A variety of crypto currencies such as Bitcoin, Darkcoin, or Peercoin (see coinmarketcap.com for a complete listing) have been in use for anonymous business transactions that are conducted within and across most Dark Web marketplaces. Hackers for hire and multilingual call centers have also accelerated the growth of Dark Web [20].

Accessing and Navigating the Dark Web

The Dark Web can be reached through decentralized, anonymized nodes on a number of networks including TOR [2] or I2P (Invisible Internet Project). TOR, which was initially released as The Onion Routing project in 2002, [16] was originally created by the US Naval Research Laboratory as a tool for anonymously communicating online. TOR "refers both to the software that you install on your computer to run TOR and the network of computers that manages TOR connections" [22]. TOR's users connect to websites "through a series of virtual tunnels rather than making a direct connection, thus allowing both organizations and individuals to share information over public networks without compromising their privacy" [2]. Users route their web traffic through other users' computers such that the traffic cannot be traced to the original user. TOR essentially establishes layers (like layers of an onion) and routes traffic through those layers to conceal users' identities.

According to Clark [22] to get from layer to layer, TOR has established "relays" on computers around the world through which information passes [22]. Information is encrypted between relays, and "all TOR traffic passes through at least three relays before it reaches its destination." The final relay is called the "exit relay," and the IP address of this relay is viewed as the source of the TOR traffic. When using TOR software, users' IP addresses remain hidden. As such, it appears that the connection to any given website "is coming from the IP address of a TOR exit relay, which can be anywhere in the world" [22]. While data on the magnitude of the Dark Web and how they relate to the Surface Web are not clear, data on TOR users do exist. According to metrics from the TOR Project, the mean number of daily TOR users in the USA across the first 3 months of 2015 was 360,775—or 16.56% of total mean daily TOR user [2]. The USA has the largest number of mean daily TOR users, followed by Germany (over 9%) and Russia (nearly 8%).

What Are the Uses of the Dark Web?

A smart person buying recreational drugs online wouldn't want to type related keywords into a regular browser. He/she will need to anonymously go online using an infrastructure that will never lead interested parties to his/her IP address or physical location [18]. Drug sellers wouldn't want to set up shop in an online location whose registrant law enforcement can easily determine or where the site's IP address exist in the real world, too. There are many other reasons, apart from buying drugs, why people would want to remain anonymous or set up sites that can't be traced back to a physical location or entity. People who want to shield their communications from government surveillance may require the cover of Dark Webs. Whistle-blowers may want to share vast amounts of insider information to journalists without leaving a paper trail. Dissidents in restrictive regimes may need anonymity in order to safely let the world know what's happening in their country [18]. On the flip side, people who want to plot the assassination of a high-profile target will want a guaranteed but untraceable means. Other illegal services like selling documents such as passports and credit cards also require an infrastructure that guarantees anonymity. The same can be said for people who leak other people's personal information like addresses and contact details.

The Impact of Anonymous Communication Networks on Cyber Security in Smart Cities

According to Cui et al. [23] smart cities are expected to improve the quality of daily life, promote sustainable development, and improve the functionality of urban systems. As many smart systems have been implemented, security and privacy issues have become a major challenge that requires effective countermeasures. However, traditional cybersecurity protection strategies cannot be applied directly to these intelligent applications because of the heterogeneity, scalability, and dynamic characteristics of smart cities [23].

Pelton and Singh [24] further noted the world of the Internet and the continuing global spread of cyber-services and new capabilities such as the Internet of Things, the cloud, and industrial control systems such as SCADA networks has expanded the scope of cyber-crime and redefined the scope of security systems in the smart city. Pelton and Singh [24] further noted that the utilization of anonymous networks such as TOR will pose huge threats to smart cities and new capabilities such as the Internet of Things.

Research Methodology

Past research mostly involved reducing the complexity of this process by first reducing the set of relays or network routers to monitor, and then identifying the actual source of anonymous traffic among network connections that are routed via this reduced set of relays or network routers to monitor. A study of various research efforts in this field reveals that there have been many more efforts to reduce the set of relays or routers to be searched than to explore methods for actually identifying an anonymous user amidst the network connections using these routers and relays. Few have tried to comprehensively study a complete attack that involves reducing the set of relays and routers to monitor and identifying the source of an anonymous connection.

The research was largely of a Design Science nature with the addition of a qualitative element. According to Hevner [25], Design science creates and evaluates IT artifacts intended to solve identified organizational problems. It involves a rigorous process to design artifacts to solve observed problems, to make research contributions, to evaluate the designs, and to communicate the results to appropriate audiences. Such artifacts may include constructs, models, methods, and instantiations [25]. They might also include social innovations or new properties of technical, social, and/or informational resources; in short, this definition includes any designed object with an embedded solution to an understood research problem.

Research Question

What is the extent of South African traffic on anonymized networks?

The study analyzed application layer header data relayed through the router to determine the protocol distribution in the anonymous network. The results showed the types of applications currently used over TOR, a substantial amount of which is non-interactive traffic.

Research Taxonomy

It is evident from Fig. 10.2 which presents a research taxonomy on various platforms, that the research platforms can be classified into three categories, namely, Experiments, Simulations, and Analysis.

Saleh [26] notes that the majority of studies undertaking TOR analysis experiment as a research (paradigm) and platform deployed their private testbeds with 1–2 clients and 1–2 servers. Several studies by Blond [27], McCoy [18], and Dingeldine [28] deployed limited number of relays for experiments. Number of clients were increased drastically in the Planet Lab and Cloud setup for TOR experiments.

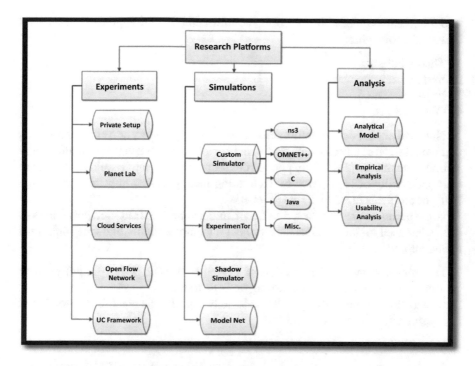

Fig. 10.2 Research taxonomies. Reproduced from: Saleh [26]

Moreover, traffic analysis was the most frequently studied topic. Majority research works used the default TOR setup without any modifications. The experiment design utilized for the study was aligned to that of McCoy [18].

Experiment Design

The research was largely of a Design Science (DS) nature, with the addition of a qualitative element. According to Hevner [25], design science creates and evaluates IT artifacts intended to solve identified organizational problems. It involves a rigorous process to design artifacts to solve observed problems, to make research contributions, to evaluate the designs, and to communicate the results to appropriate audiences. Such artifacts may include constructs, models, methods, and instantiations [25]. They might also include social innovations or new properties of technical, social, and/or informational resources; in short, this definition includes any designed object with an embedded solution to an understood research problem.

Three papers from the early 1990s introduced DS research to the IS community. According to Nunamaker [29] who advocated the integration of system development into the research process, by proposing a multimethod logical approach that would include the following:

1. Theory building.
2. Systems development.
3. Experimentation.
4. Observations.

Nunamaker [29] defined information systems design theory as a class of research that would stand as an equal with traditional social science based theory building and testing. Nunamaker [29] further pointed out that design research could contribute to the applicability of IS research by facilitating its application to better address the kinds of problems faced by IS practitioners.

An experiment was conducted to better understand the TOR Network and how it is being used in South Africa. The project encompassed the following high-level methodology:

1. Development and implementation of customized TOR node (bridge) code to allow for the logging and monitoring of TOR traffic.
2. Connection to the existing TOR node network and enable full control based logging of the following events:

   (a) *CIRC*

   Information on newly created, already existing and closed TOR nodes. This module is responsible for ensuring the connection to the TOR network is maintained.

   (b) *STREAM*

   Information on status of application streams including which circuit is used for the connection (e.g.,, HTTP-based connection data).

   (c) *ORCONN*

   Newly established and closed connections to TOR nodes.

   (d) *BW*

   Bandwidth utilized by the TOR node.

   (e) *STREAM_BW*

   Bandwidth used by the various streams within the TOR node configuration.

   (f) *DEBUG, INFO, NOTICE, WARN, ERR*

   Information messages related to the running of the TOR node.

   (g) *ADDRMAP*

   Domain-to-IP mapping that is cached by TOR client to determine the actual Internet address of the connected client.

(h) *NEWDESC, AUTHDIR_NEWDESCS, DESCCHANGED*
    TOR directory services.

3. Implementation of geo based firewall rules to prevent connecting for Internet IP addresses outside of the geographical border of South Africa.
4. Collect and analyze logged data.

Configuration of a Private TOR Network

A small, stand-alone test network was created to conduct experiments on TOR. This was done so as to not disturb any nodes that were functioning as part of the live global TOR network operating on the Internet.

Network Layout

The network consists of a TOR exit node and the exit node was running the latest version of TOR. The hardware components for the TOR exit node configuration enlisted the following:

- Front end Dark web pair routing devices (16 Gig Ram, 256 Gig SSD, 4 TB attached RAID, Secondary NIC (Software MAC allocation)
- Client Management System (8 GB RAM, 512 GB HDD)

Figure 10.3 shows a graphical representation of the layout of the network. The monitors represent the end users of TOR; the red hexagon symbol is the TOR exit node. In order to access content on the TOR network a user will bypass the exit node, allowing us to capture their IP address and gain valuable information into the websites visited by the user.

TOR Configuration

TOR determines all of its configuration settings from a file called the Torrc file. Settings listed in this file tell TOR what name to use, what services to run, what types of logging to perform, what policies to enforce, and, for the stand-alone environment, which directory servers to use. The specification of these is important because the relays must know where to upload their descriptors and the clients must know where they can find a list of relays to use. By default, TOR uses the directory servers that have been hard-coded into it. By providing a list of the own directory servers, the study informs TOR to use one of ours and to not use any of the preloaded ones.

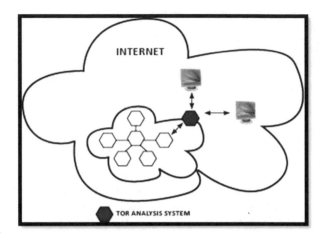

Data Collection Methodology

To better understand South African TOR usage, the study sets up a TOR exit node
on a 5 GB/s network link. This node joined the currently deployed network during
the period of 27th September–5th November 2017. The first week was utilized as
a testing period for the exit node and the data gathered during this period was
not utilized in the analysis of this study. During the first week of the pilot study
there were no configuration issues experienced on the server. The first 3 days of
the testing period, a minimal amount of bandwidth was allocated to the TOR exit
node. This bandwidth was increased over the forthcoming days of the testing period.
An increase in the bandwidth allocation to the exit node saw excessive amounts of
traffic passing through the node; however, the node remained stable. The only issue
encountered over the testing period were the cyberattacks the node came under.

This configuration allowed the researcher to record a large amount of TOR traffic
in short periods of time.

Results

TOR Usage

As reflected in Fig. 10.4, the average number of daily South African TOR users
ranged between 4000 and 6000 between the month of October and November 2017.

The experiment was conducted during the last week of September 2017 and
ending during the first week of November 2017; thus, the contribution of TOR
usage of South Africans is therefore small in comparison to the total number of
TOR users globally. During the period of the experiment as mentioned above, the

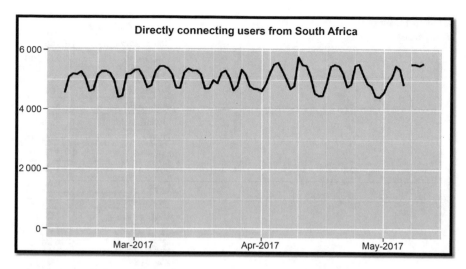

Fig. 10.4 Directly connecting users in South Africa. Source: TOR [30]

total number of TOR users globally range between 275,000 and 300,000 users and South Africa contributed only 2% of the total number of TOR users globally, this is disproportionately small to countries such as USA and Germany.

Final Observation

The experiment yielded a total of 11,763,480 hits on various websites, which were thereafter categorized into 52 different categories which varied from social media to the visiting of extreme websites such as pornography, weapons, and hacking as shown in Table 10.1.

The largest proportion of websites visited during the experiment was classified as social media sites. According to Statistics South Africa [31] 70% of South Africans weekly activities are spent online, visiting social media sites. As a large proportion of traffic passing through the exit node as shown in Fig. 10.5 was during working hours one could ascertain that employers were utilizing TOR at work to gain access to social media sites.

The primary usage of TOR in a South African context is used to bypass a company's proxy server to allow for anonymous browsing. This allows employees at organizations to access social networking sites and search for jobs without being traced on a company's network. This places a huge risk not only to the companies' network but also for individuals' accessing social networks on TOR. Studies by Flores [32], Django [33], and Dingeldine [34], whilst monitoring traffic on the Dark Web found the predominant usage of TOR was to access illicit websites and engage in unlawful trade. The findings presented in Fig. 10.5 do not support the studies

Table 10.1 Website classification

URL classification	Hits
Social networks	2,208,750
Job search	718,900
Web TV	651,326
Advertising	567,840
Web radio	519,586
Update sites	389,474
Porn	285,348
Spyware	277,695
Sex	180,235
Forum	169,024
Automobile	163,080
URL shortner	158,928
Tracker	157,724
Government	157,718
Alcohol	156,536
Hacking	153,504
Radio/TV	151,008
ISP	150,282
Homestyle	148,720
Music	147,828
Webmail	147,318
Web phones	144,823
Warez	144,807
Drugs	143,360
Military	142,740
Chat	140,990
Downloads	140,685
Science	139,113
Finance	139,000
Ringtones	138,320
News	137,718
Religion	136,956
Dating	136,398
AnonVPN	135,300
Pod casts	134,536
Models	133,750
Gamble	133,358
Recreation	133,342
Aggressive	132,990
Image hosting	132,225
Search engines	129,870
Politics	129,840

(continued)

Table 10.1   (continued)

URL classification	Hits
Remote control	127,534
Violence	124,440
Hospitals	124,230
Shopping	123,903
Education	123,615
Hobby	122,089
Library	121,847
Movies	121,555
Weapons	118,215
Dynamic	115,107
Total URL hits	11,763,480

Reproduced   by: Researchers Development

The predominant usage of TOR by South Africans is depicted in this table

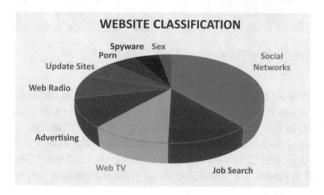

Fig. 10.5 Website classification. Source: Researchers Development

presented by Blond [27], Chertoff [35], and Dingeldine [34]. A potential flaw in their studies was that they only monitored exit routing traffic to countries such as Germany, the USA, and parts of Europe. Vitaris [36] noted that the majority the majority of pornographic websites and drug trade could be found in countries such as the USA, Germany, and Europe.

Figure 10.6 provides a further declassification of Social media ebsites.

A declassification of the social media websites shows that the majority of the social media traffic was directed to Facebook. Facebook has recently launched a TOR browser version of their social media platform. Django's [33] observations of Dark Web traffic found the dominant amount of traffic passing through the exit node was that off malicious activities. The primary use of TOR in Django [33] study was the usage of the Dark Web to purchase drugs and to visit pornographic websites. A further study by Flores [32] only noted the use of emails by West Africans when

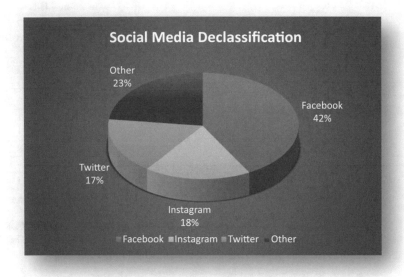

Fig. 10.6 Social media distribution. Source: Researchers Development

targeting individuals and businesses. Flores [32] findings in terms of social media activities on the Dark Web were only related to scam emails and a few chat groups. The findings from this study do not correlate with that off Django [33] and Flores [32], as they noted the major use of TOR was to access pornographic websites, illicit trade, and the use of TOR to distribute scam emails on the Dark Web.

Figure 10.7 represents a classification of the chat activity associated with the social media activities.

There was a total of 136,398 visits to dating websites. Findings from Flores [32] noted the use of TOR by West Africans when targeting individuals and businesses with scam activities. Flores [32] further noted the use of TOR by West Africans in developing scam emails and posing fraudulent profiles on social media websites. This study, however, did not log any traffic to any off the West African countries. This study therefore proves the use of TOR by South Africans on these dating websites to be associated with scam activity and remaining anonymous in doing so. This finding will therefore add to the findings to that of Flores [32] where the issue of the utilization of TOR in scamming individuals and businesses is not only confined to West African countries but also associated to South Africa as well.

This will also tie in with theoretical framework of the study that is based on the Space Transition Theory as proposed by Jaishankar [37], where he explains the behaviors of the persons who bring out their conforming and nonconforming behaviors in the physical space and virtual space. Jaishankar [37] also noted that

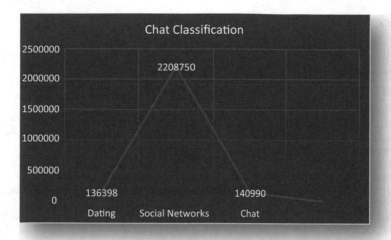

Fig. 10.7 Chat classification. Source: Researchers Development

in the Space Transition Theory people behave differently when they move from one space to another. Individuals that repress crime in a physical space, would commit such crimes in the virtual space. Jaishankar [37] also noted concepts such as cyberstalking and cyber-defamation where offenders make use of online space because of its anonymity and widespread approach.

Conclusion

The primary objective of the study was to understand TOR usage by South Africans. In particular, the study provided observations that helped understand how TOR is being used, how TOR is being misused, and who participates in the network as clients and routers. Given the ease at which an eavesdropping exit router can log sensitive user information, a method for detecting malicious logging of exit routers was developed and evidence provided that there are such routers that specifically log insecure protocol exit traffic. In collaboration with changes in the South African legal landscape (implementation of laws and structures to govern South African Internet use), it is likely that this study will provide an independent analysis and justification for some of the controls proposed.

References

1. Caverlee D. (2015). The deep web and dark the darknet. *IEEE Transactions on Services Computing, 2*(4), 5–8.
2. TOR. (2016). *TOR metrics portal.* Retrieved from TOR Project: http://www.torproject.org/
3. Adler, M. W. (2012). An analysis of the degradation of anonymous protocols. In *Network and Distributed Security Symposium.*
4. Bauer, K. (2007). Low resource routing attacks against TOR. In *ACM Workshop* (pp. 11–20).
5. Burch, B. (2013). Tracing anonymous packets to their approximate source. In *USENIX Conference on System Administration* (pp. 319–328). LISA.
6. Murdoch, S. (2012). Hot or not: Revealing hidden services by their clock skew. In *ACM Conference* (pp. 27–36).
7. Edman, E. (2012). AS-awareness in TOR path selection. *Journal of Information Management, 2*(1), 380–389.
8. Feamster, N. (2012). *Location diversity in anonymity networks* (pp. 66–76). New York: ACM.
9. Johnson, A. (2013). Users get routed: Traffic correlation on TOR by realistic adversaries. *Journal of Information Management, 8*(1), 121–131.
10. Murdoch, M. (2015). *Low-cost traffic analysis of TOR. s.l* (pp. 183–195). New York: ACM.
11. Shimaikov, V. (2011). *Timing analysis in low latency mix networks: Attacks and defences* (pp. 121–134). Piscataway, NJ: IEEE.
12. Wright, M. (2012). *An analysis of the degradation of anonymous protocols* (pp. 116–122). MIS.
13. Pappas, V. (2010). Compromising anonymity using packet spinning. *Journal of Information Management, 2*(3), 161–174.
14. Ovelier, L. (2006). *Locating hidden servers* (pp. 161–174). New York: ACM.
15. Evans, N. (2011). A practical congestion attack on TOR using long paths. *Journal of MIS, 6*(6), 33–50.
16. Dingeldine, R. (2010). TOR: The second-generation onion router. *Journal of Management Information Systems, 18*(8), 303–319.
17. Reiter, M. (2012). *Crowds: Anonymity for web transactions* (pp. 66–92). New York: ACM.
18. McCoy, D. (2008). Shining light in dark places. Understanding the TOR network. In *8th International symposium on Privacy Enhancing Technologies* (pp. 63–76).
19. Isdals, T. (2010). Privacy-preserving P2P data sharing with Oneswarm. *Springer Journal of Information Security and Privacy, 5*(7), 111–122.
20. Ablon, L. (2014). *Markets for cybercrime tools and stolen data.* Santa Monica, CA: Rand Corporation.
21. Bartlett, J. (2015). *The dark web: Inside the digital underworld.* London: William Heinemann Ltd.
22. Clark, A. (2013). *TOR. The anonymous internet, and if it's right for you.* Accessed in Gizmodo (Vol. 1_2, pp. 112–113).
23. Cui, L. X., Qu, Y., Gao, L., & Yang, Y. (2018). Security and privacy in smart cities: Challenges and opportunities. *IEEE Access, 6,* 46134–46145.
24. Pelton, J. N., & Singh, I. B. (2019). Coping with the dark web, cyber-criminals and techno-terrorists in a smart city. In *Smart cities of today and tomorrow.* Cham: Copernicus. ISBN 978-3-319-95821-7.
25. Hevner, A. (2004). Design research in information systems research. *Journal of Management Information Systems, 3*(6), 75–105.
26. Saleh, S. (2017). *Shedding light on the dark corners of the internet: A survey of TOR research* (pp. 1–35). Piscataway, NJ: IEEE.
27. Blond, L. (2012). One bad apple spoils the bunch: Exploiting P2P applications to trace and profile TOR users. *ARXIV* (pp. 15–18).
28. Dingeldine, R. (2013). TOR. The second generation onion router. In *NISEX security* (pp. 33–52).

29. Nunamaker, J. (1991). Systems development in information systems research. *Journal of Management Information Systems, 5*(1), 89–106.
30. TOR Project. (2018). *Ethical TOR research: Guidelines*. https://blog.torproject.org/blog/ethical-tor-research-guidelines
31. Statistics South Africa. (2017). The many faces of cybercrime. *Trend Micro Security News, 3*(3), 65–76.
32. Flores, R. (2016). Sextortion in the far east. *Trend Micro Security News, 1*(1), 4–8.
33. Django, T. (2018, 30 August). *TOR. The anonymous internet, and if it's right for you* (pp. 112–113). New York: Gizmodo.
34. Dingeldine, R. (2015). The second-generation onion router. *Springer Journal of Information Security and Privacy, 4*(1), 4–8.
35. Chertoff, M. (2015). The impact of the dark web on internet governance and cyber security. *Global Commission on Internet Governance* (Vol. 1_2, pp. 101–105).
36. Vitaris, B. (2016). *Russian is collecting encryption keys as Änti-Terrorism legislation goes into effect*. [Online] Available at: Retrieved May 21, 2018, from https://www.deepdotweb.com/2016/08/03/russian-collecting-encryption-keys-anti-terrorism-legislation-goes-effect/
37. Jaishankar, K. (2007). Establishing a theory of cyber crime. *International Journal of Cyber Criminology, 1*, 7–9.

Chapter 11
LBCLCT: Location Based Cross Language Cipher Technique

Vishal Gupta, Rahul Johari, Kalpana Gupta, Riya Bhatia, and Samridhi Seth

Nomenclature

IaaS	Infrastructure as a Service
PaaS	Platform as a Service
SaaS	Software as a Service
IT	Information Technology
AES	Advanced Encryption Standard
HMAC	Hash Message Authentication Code
SDK	Software Development Kit
IDE	Integrated Development Environment

(SWINGER (swinger@ipu.ac.in): Security, Wireless, IoT Network Group of Engineering and Research) Lab Members

V. Gupta · R. Johari (✉) · K. Gupta · R. Bhatia · S. Seth
University School of Information, Communication & Technology (USICT), Guru Gobind Singh Indraprastha University, Dwarka, Delhi, India

Center for Development of Advanced Computing, Noida, India
e-mail: rahul@ipu.ac.in; swinger@ipu.ac.in

© Springer Nature Switzerland AG 2020   221
F. Al-Turjman (ed.), *Smart Cities Performability, Cognition, & Security*,
EAI/Springer Innovations in Communication and Computing,
https://doi.org/10.1007/978-3-030-14718-1_11

Introduction

Cloud computing is a model for enabling convenient, on demand network access to a shared pool of configurable computing resources (e.g., networks, servers, storage, applications, and services) that can be rapidly provisioned and released with minimal management effort or service provider interaction [1]. Service models existing in cloud computing are detailed as follows.

Infrastructure as a Service

IaaS is where physical server space is rented and kept at a vendor's data warehouse. The customer can install any legal software to the server that enables him to allow access to his staff and clients. The IaaS layer offers storage and computer resources that developers and IT ions can use to deliver business solutions [2].

Platform as a Service

PaaS is where the operating system is hosted "in the cloud," rather than being physically installed on end user hardware. The PaaS layer offers standard remote services with which developers can build applications on top of the computer infrastructure. This might include developer tools that are offered as a service to access data, build other services such as billing, database, etc. [3].

Software as a Service

For SaaS, the service provider hosts the software so end user doesn't need to install it, manage it, or buy hardware for it. End user has to establish, connect, and use it. SaaS examples include customer relationship management as a service, email, logistics software, order management software, payroll software, and any other software which is hosted on the Internet and not physically installed on end user computer. Software as a service SaaS is where most businesses start their journey to cloud computing, typically starting with the remote delivery of email and online backup of business information [4].

In the current chapter, a mobile bill payment application has been designed and developed in Java programming language for the purpose of secure bill payment over the cloud. The web application is hosted on cloud. Google Cloud Platform, "Google App Engine," is used for the deployment of mobile bill payment application. A financial application that contains confidential information needs to be secure. Affine cipher technique is used with geographic coordinates to encrypt the information and compare with rail fence cipher technique. After encryption, language translation has been done on cipher text.

To the best of our knowledge, this is first of its kind of work in this direction. For the sake of convenience and simplicity, the rest of the paper is organized as follows: Sect. 11 describes deployment models of cloud computing, Sect. 11 describes about cryptography, Sect. 11 describes literature survey, Sect. 11 describes the methodology adopted, Sects. 11 and 11 describe the encryption using affine cipher and encryption using rail fence cipher, respectively, Sect. 11 describes translation mapping, Sect. 11 describes algorithm/pseudocode, Sect. 11 describes the simulation environment and shows the flow chart of the proposed work, and Sect. 11 shows the results followed by conclusion and references.

Deployment Models for Cloud Architecture Solution

Private Cloud

It refers to cloud computing on private networks. Private clouds are built for the use of one client, providing full control over data, security, and quality of service. Private clouds can be built and managed by a company's own IT organization or by a cloud provider.

Community Cloud

The cloud infrastructure is used by several organizations and helps a particular community that has communal concerns. It may be managed by a third party or the organization.

Public Cloud

Computing resources are available on the Internet through the web applications or web services at run time from third-party provider. Public clouds are widely used by third parties, and applications from different customers are likely to be mixed together on the cloud's servers, storage systems, and networks.

Hybrid Cloud

A hybrid cloud environment combines multiple public and private cloud models. Hybrid clouds introduce the complexity of determining how to distribute applications across both a public and private cloud [5].

Cryptography

In today's time, everyone in some context is attached to Internet and the system attached to the Internet is under influence from viruses and various attacks from the attacker. User's information may be sensitive such as bank account information, insurance policy details, Aadhaar number, Permanent account number, credit card detail, and biometric details, which cannot be shared with any one on the network. Therefore nowadays, it is necessary to protect the information from suspicious activities that may misuse or harm the user's information.

Cryptography is the way to protect information that is to be sent over the network. The original information is converted into some other form which is not understandable by others called cipher text and this cipher text is transmitted over the network. This process of converting original data into cipher text is called encryption. At the receiver side, the cipher text is again converted into original form of data. This process is called decryption.

Literature Survey

In [6] the authors discussed the encryption technique that improved substitution approach to encrypt the plain text and which uses poly-alphabetic cipher technique. Hill cipher technique is applied to the encrypted text for generating the new cipher text. This double process is used to make encryption approach more strong and secure than the previous techniques. In [7] the authors discussed about the deceitful activities such as financial fraud, credit card fraud transaction, and insurance fraud which are of important concerns to many organizations including insurance companies, banks, and public service organizations. The authors identified some issues which are not solved including real time fraud detection and discussed the fraud data analytics techniques with big data. In [8] the authors described the web application security threats that posed some challenges for data privacy and security. Major vulnerabilities that are part of Open Web Application Security Project have been described. A tool such as Burp Suite [9] is explained for how to detect and analyze the vulnerabilities. The authors explained the secure development life cycle to mitigate the vulnerabilities. Detection of web threats with the help of security testing and penetration testing is also discussed. In [10] the authors discussed about the one time pad cipher encryption technique. The one time pad cipher's key is generated using Fibonacci series formula. To make the generated cipher more difficult to decrypt, the affine cipher was integrated with one time pad cipher encryption technique. The authors used Borland Delphi 7.0 to make encryption and decryption easier. In [11] the author described about the cryptography which is an important matter of security and confidentiality of communication over Internet. The author introduced the classic and modern encryption. As a classic encryption, affine cipher is analyzed and improved by using extended set of alphabet, encryption of packet, and hash function. In [12] the authors introduced the concept of online voting using the voter's biometrics. An application is designed and developed using Android SDK. The application has aimed to vote through digital medium such as Android mobile platform. In [13] the authors presented a topic model to access the activities of an individual. Geo-location information is used from social media for the classification of activities of individual. Many activity patterns have been found from individual activity data through social media. The authors extended this model to observe user specific patterns and also extended to account for missing activities. In [14] the authors described the deployment of web application on cloud platform. A financial transaction web application is designed and developed in Java programming language. The web application is deployed on Google App Engine. In [15] the authors discussed about the security of data over the network. Various cryptographic attacks have been described. Some network attacks such as brute force attack and dictionary attack have been demonstrated on the policy premium payment web application that is hosted on cloud platform. In [16] the author discussed the security issues in the service models (SaaS, PaaS, and IaaS) of the cloud computing. The author presented a qualitative analysis of threats and vulnerabilities towards service models. In [17] the authors discussed about the secure search and recovery of user data in cloud framework. A new approach of dual encryption has been introduced to provide more security to the current techniques of fuzzy keyword search. For improvement in information security, the authors used symmetric and asymmetric algorithms. In [18] the authors described about the dynamic allocation of virtual machine in cloud computing. Utilization threshold methodology is used which provides absolute median deviation for setting up threshold. In [19] the authors proposed a new cryptographic technique which is simple yet powerful. The authors also showed the effectiveness of technique by comparing it with well-known classical cipher techniques. In [20] the authors came up with a triplicative encryption scheme comprising AES standard to ensure the confidentiality, integrity, and authenticity. The authors also implemented the technique on alphabetical, numerical, and alphanumerical data. In [21] the authors combined the Caesar cipher and rail fence cipher techniques with a cross language cipher technique and implemented them in Java. The authors simulated and compared the results of these techniques. In [22] the authors introduced a framework to ensure authentication, confidentiality, and integrity of data in wireless sensor networks. The authors aimed to propose encryption by using elliptic curve cryptography. In [23] the authors focused on improvising the task scheduling in homogeneous as well as heterogeneous cloud-based IoT applications. Algorithms using swarm optimization have been implemented and results have been compared for different experienced data traffic categories. In [24] the authors proposed an industry-oriented method of multipath routing in industrial Internet of things (IIoTs). The technique is tested and results are compared with that of the standard optimization techniques. In [25] the authors came up with various local search algorithms to solve the issue of limited energy of sensors in wireless sensor networks (WSNs) when the initial energy of sensors varies. The three proposed algorithms are simulated and the result is compared with existing dominating set algorithm. In [26] the authors introduced a framework to establish a secure and seamless mutual authentication in 5G network. Hash and global assertion value have been used to show how the system achieves security goals of wireless medical sensor networks (WMSNs) in short time. In [27] the authors designed and assessed a network security model to deal with the increasing confidentiality and security threats in Internet of things (IoT). Various components as well as fundamentals of network security have also been discussed.

Methodology Adopted

In the current chapter to show the effective working of the proposed language translation technique, it has been encrypted twice, first by using key based encryption—"Affine Cipher Technique," then by using keyless cryptographic technique—"Rail Fence Cipher Technique."

Encryption Using Affine Cipher

Encryption techniques have been used to secure information before sending over the network. A mobile bill payment application has been designed and developed. User's confidential information has been encrypted using affine cipher encryption technique. Affine cipher, a classic cryptographic approach, has been used in this work to encrypt the sensitive information of the user on the network. An improvement has been made to the classic affine cipher technique by introducing one dynamic key instead of static key. The concept of geo-location has been used for generating the keys for the affine cipher technique. The keys are obtained by the geographic coordinates of the user, which are latitude and longitude. The first key is kept static and second key is dynamic whose value is the sum of latitude and longitude value. The improved affine cipher technique makes the information more strong and secure.

Encryption	$E(x) = (key1 * x + key2) \bmod 26$
Decryption	$D(x) = (inverse(key1) * (x - key2)) \bmod 26$

Encryption Using Rail Fence Cipher

For sending the important information over the network, it needs to be secure before transmitted. User's important information has been encrypted using rail fence cipher technique. Rail fence is a transposition cipher technique. This technique involves the following steps: first, writing the plain text elements as a sequence of diagonal and second, to generate cipher text, read it as sequence of row. For example, suppose a plain text is: encryption. The plain text would be written as:

e		c		y		t		o	
	n		r		p		i		n

To encrypt, cipher text is produced by reading across the rows. The cipher text is: ecytonrpin.

Translation Mapping

Cipher text has to be stronger before sending over the Internet so that it becomes impossible or it may take a long time to decrypt for an attacker. Cipher text which is generated after the improved affine encryption technique is translated into two Indian languages. First, original cipher text which is in the form of digits has been translated into English language alphabets. Second, the mapped English language characters are further translated into Hindi language characters. This process made the original confidential information more secure.

Proposed Algorithms/Pseudocode

Two algorithms are hereby proposed to handle the credit card security by validating the user information and validating credit card number. Figure 11.1 depicts the flowchart for the proposed algorithm.

Algorithm 1 Validating user data

Notation
mnum: mobile number
bamt: bill amount
cnum: credit card number
cvv: card verification value
edate: expiry date
Trigger: User submit the Plain Text
if (mnum=NaN or length(mnum)!=10)
        print "mobile number must be of 10 digits"
elseif (bamt=NaN)
        print "please enter valid amount"
elseif(cnum=NaN or length(cnum) !=16 or!(luhnCheck(cnum)))
        print "card number is not valid"
elseif (cvv=NaN or length(cvv)!=3)
        print "cvv number must be of 3 digits"
elseif (edate<december,2018)
        print "value must be december,2018 or later."
elseif (edate>december,2021)
        print "value must be december,2021 or earlier."
else
        print "successful"

Algorithm 2 Validating credit card number using Luhn Mod10 algorithm [28]

Trigger: credit card number (cnum) validation
len: length of credit card number, total: 0
luhnCheck(cnum) {
for i from 0 to len -1
        digit: cnum[i]
        if (i%2==0)
                digit = digit*2
        if (digit < 9)
                digit = digit /10 + digit % 10
        total = total + digit
if (total % 10==0)
        print "Valid"
else
        print "Card number is not valid"
}

Fig. 11.1 Flow chart
showing encryption process

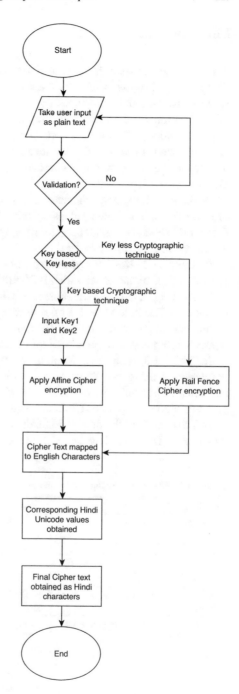

Simulation and Result

The proposed system is simulated on Google platform, Google App Engine. Google App Engine is chosen over other cloud platforms because it supports several programming languages and facilitates the development and deployment of application on Google Cloud Server. In terms of hardware requirements, for implementation, a 32 bit Intel core i3 processor @ 1.90 GHz, RAM being 4 GB in size was used. In terms of software requirements, for implementation, Microsoft Windows operating system (windows 7), Eclipse IDE Mars Version, and Java Development Kit 1.7 were used.

In Table 11.1, a special character (.) is used as a separator during affine encryption to separate the encrypted values so that it might be used in decryption process. Further this character is mapped to other special character (?) when encrypted data is mapped to English characters.

In Table 11.1, Key1 is considered to be static with value 5 as value of this key and size of the alphabet set which is 26 must be co-prime.

In Table 11.2, at the destination side, Cipher text (in Hindi language) is received by the user. The unicode value of each received Hindi character was then translated or mapped to its corresponding English language character. Thereafter the affine cipher technique was applied to obtain back the original credit card number.

In Table 11.3, at source side, plain text (in English) that is the credit card number is encrypted using rail fence cipher which is mapped to corresponding English characters. Then unicode values are found for each of the characters and the values are translated to respective Hindi characters to be sent as final cipher text.

In Table 11.4, at the destination side, cipher text (in Hindi language) is received by the user. The unicode value of each received Hindi character was then translated

Table 11.1 Example showing encryption process using key based technique (affine cipher)

Process	Text
User input as plain text	5214789741236547 (sample 16 digit credit card number)
Affine cipher encrypted text Key1=5(Static) Key2=105 (longitude + latitude)	0.11.6.21.10.15.20.10.21.6.11.16.5.0.21.10
Cipher text mapped to English characters	A?BB?G?CB?BA?BF?CA?BA?CB?G?BB ?BG?F?A?CB?BA?
Corresponding Hindi language unicode values	2309,2404,2348,2348,2404,2327,2404,233 0,2348,2404,2348,2309,2404,2348,2347,2 404,2330,2309,2404,2348,2309,2404,2330 ,2348,2404,2327,2404,2348,2348,2404,23 48,2327,2404,2347,2404,2309,2404,2330, 2348,2404,2348,2309,2404
Final cipher text to be transmitted as Hindi characters	अ।बब।ग।चब।बअ।बफ।चअ।बअ।चब।ग।बब। बग।फ।अ।चब।बअ।

Table 11.2 Example showing decryption process using key based technique (affine cipher)

Process	Text
Cipher text received	अ।बब।ग।चब।बअ।बफ।चअ।बअ।चब।ग।बब। बग।फ।अ।चब।बअ।
Unicode value for the cipher text received	2309,2404,2348,2348,2404,2327,2404,233 0,2348,2404,2348,2309,2404,2348,2347,2 404,2330,2309,2404,2348,2309,2404,2330 ,2348,2404,2327,2404,2348,2348,2404,23 48,2327,2404,2347,2404,2309,2404,2330, 2348,2404,2348,2309,2404
Unicode value mapped to English characters	A?BB?G?CB?BA?BF?CA?BA?CB?G?BB ?BG?F?A?CB?BA?
Text after language translation	0.11.6.21.10.15.20.10.21.6.11.16.5.0.21.10
Affine cipher decrypted text using two keys	5214789741236547

Table 11.3 Example showing encryption process using key less technique (rail fence cipher)

Process	Text
User input as plain text	5214789741236547 (sample 16 digit credit card number)
Rail fence cipher encrypted text	5179426424871357
Cipher text mapped to English characters	FBHJECGECEIHBDFH
Corresponding Hindi language unicode values	2347,2348,2361,2332,2311,2330,2327,231 1,2330,2311,2312,2361,2348,2342,2347,2 361
Final cipher text to be transmitted as Hindi characters	

Table 11.4 Example showing decryption process using key less technique (rail fence cipher)

Process	Text
Cipher text received	फबहजइचगइचइईहबदफह
Unicode value for the cipher text received	2347,2348,2361,2332,2311,2330,2327,231 1,2330,2311,2312,2361,2348,2342,2347,2 361
Unicode value mapped to english characters	FBHJECGECEIHBDFH
Text after language translation	5179426424871357
Rail fence cipher decrypted text	5214789741236547

Table 11.5 Comparison of cryptographic techniques on the basis of line of code and execution time

Cryptographic technique	Affine cipher	Rail fence cipher
Line of code (LOC)	121	117
Execution time (ns)	3,272,977	535,376

or mapped to its corresponding English language character. Thereafter the rail fence cipher technique was applied to obtain back the original credit card number.

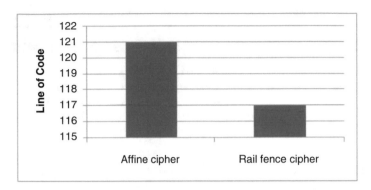

Fig. 11.2 Graphical representation of comparison of affine cipher and rail fence cipher on the basis of line of code

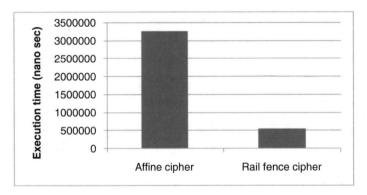

Fig. 11.3 Graphical representation of comparison of affine cipher and rail fence cipher on the basis of execution time

The above graphs depict that although there is not much difference in the lines of code executed for both techniques but affine cipher technique took at least 6 times more execution time as compared to rail fence cipher. To demonstrate the effectiveness of the proposed algorithm, comparision has been made on the basis of two parameters: execution time and line of codes. Table 11.5 depicts the data in tabular form. Figures 11.2 and 11.3 graphically depict the comparison.

Conclusion and Future Scope

The task of design, development, and deployment of mobile bill payment application in J2EE was successfully achieved. For wider reach for the open source community, the application is hosted on the web server: Google App Engine, which is a platform as a service. In this paper two encryption techniques, one is key based with geographic coordinates as key and other is keyless, were implemented along with cross language mapping to generate cipher text. In future, representation of same work with more techniques such as Vigenere cipher, Hill cipher, Caesar cipher, simple substitution cipher, and Playfair cipher would be implemented. Besides latitude and longitude, other set of keys such as IP (Internet Protocol) address, MAC (Media Access Control) address of device and social credentials of an individual like driving license, voter ID, PAN number, and AADHAR card number is also proposed to be taken. Apart from that, applying encryption in Hindi text is also proposed. Our proposed algorithm of Location Based Cross Language Cipher Technique (LBCLCT) can be integrated in AES/DES algorithm to increase the level of security.

Submission and Acknowledgement   This chapter is an expanded and extended version of CLCT [29]. The authors wish to acknowledge the research oriented environment provided by GGSIP University.

References

1. Brown, E. (2011). Final version of NIST cloud computing definition published. Diambil dari. Retrieved October 10, 2014. http://www.nist.gov/itl/csd/cloud-102511
2. Zia, A., & Khan, M. N. A. (2012). Identifying key challenges in performance issues in cloud computing. *International Journal of Modern Education and Computer Science, 4*(10), 59.
3. Ahamed, F., Shahrestani, S., & Ginige, A. (2013). Cloud computing: Security and reliability issues. *Communications of the IBIMA, 2013*, 1–12.
4. Kaur, G., & Chawla, S. (2014). Cloud computing for business: Models and platforms. *International Journal of Computer Science and Engineering, 1*(7).
5. Sriram, I., & Khajeh-Hosseini, A. (2010). Research agenda in cloud technologies. arXiv:1001.3259.
6. Rajput, Y., Naik, D., & Mane, C. (2014). An improved cryptographic technique to encrypt text using double encryption. *International Journal of Computer Applications, 86*(6), 24–28.
7. Makki, S., Haque, R., Taher, Y., Assaghir, Z., Ditzler, G., Hacid, M.S., et al. (2017). Fraud data analytics tools and techniques in big data era. In *2017 International Conference on Cloud and Autonomic Computing (ICCAC)* (pp. 186–187). Piscataway: IEEE.
8. Deshpande, V. M., Nair, D. M. K., & Shah, D. (under review, 2017). Major web application threats for data privacy & security–detection, analysis and mitigation strategies. *International Journal of Scientific Research in Science and Technology*. PRINT ISSN: 2395–6011.
9. Joshi, C., & Singh, U. K. (2016). Security testing and assessment of vulnerability scanners in quest of current information security landscape. *International Journal of Computer Applications, 145*(2), 1–7.
10. Firdaus, I. L., Marwati, R., & Sispiyati, R. (2018). Aplikasi kriptografi komposisi one time pad cipher dan affine cipher. *Journal EurekaMatika, 5*(2), 42–51.
11. Wu, Y. (2015). Improvement research based on affine encryption algorithm. In *2015 14th International Symposium on Distributed Computing and Applications for Business Engineering and Science (DCABES)* (pp. 167–170). Piscataway: IEEE.
12. Mahajan, M., Wagh, M., Biswas, P., Alai, S., & More, S. (2017). M-vote (online voting system). *International Journal, 2*(10), 10–14.
13. Hasan, S., & Ukkusuri, S. V. (2014). Urban activity pattern classification using topic models from online geo-location data. *Transportation Research Part C: Emerging Technologies, 44*, 363–381.

14. Vishal, K., & Johari, R. (2017). CBFT: Cloud based financial transaction application. In *2017 6th International Conference on System Modeling & Advancement in Research Trends-SMART*. Piscataway: IEEE.
15. Vishal, K., & Johari, R. (2018). SOAiCE: Simulation of attacks on cloud computing environment. In *2018 8th International Conference on Cloud Computing, Data Science & Engineering-Confluence*. Piscataway: IEEE.
16. Singh, A. (2018). Security concerns and countermeasures in cloud computing: a qualitative analysis. *International Journal of Information Technology, 3*, 1–8.
17. Tariq, H., & Agarwal, P. (2018). Secure keyword search using dual encryption in cloud computing. *International Journal of Information Technology, 3*, 1–10.
18. Kumar, M., Yadav, A. K., Khatri, P., & Raw, R. S. (2018). Global host allocation policy for virtual machine in cloud computing. *International Journal of Information Technology, 10*(3), 279–287.
19. Gupta, S., Johari, R., Garg, P., & Gupta, K. (2018). C 3 T: Cloud based cyclic cryptographic technique and it's comparative analysis with classical cipher techniques. In *2018 5th International Conference on Signal Processing and Integrated Networks (SPIN)* (pp. 332–337). Piscataway: IEEE.
20. Johari, R., Bhatia, H., Singh, S., & Chauhan, M. (2016). Triplicative cipher technique. *Procedia Computer Science, 78*, 217–223.
21. Kumar, S., Johari, R., Singh, L., & Gupta, K. (2017). SCLCT: Secured cross language cipher technique. In *2017 International Conference on Computing, Communication and Automation (ICCCA)* (pp. 545–550). Piscataway: IEEE.
22. Al-Turjman, F., & Alturjman, S. (2018). Confidential smart-sensing framework in the IoT era. *The Journal of Supercomputing, 74*(10), 5187–5198.
23. Al-Turjman, F., Hasan, M. Z., & Al-Rizzo, H. (2018). Task scheduling in cloud-based survivability applications using swarm optimization in IoT. *Transactions on Emerging Telecommunications*. https://doi.org/10.1002/ett.3539
24. Al-Turjman, F., & Alturjman, S. (2018). 5G/IoT-enabled UAVs for multimedia delivery in industry-oriented applications. *Multimedia Tools and Applications, 78*, 1–22.
25. Pino, T., Choudhury, S., & Al-Turjman, F. (2018). Dominating set algorithms for wireless sensor networks survivability. *IEEE Access, 6*, 17527–17532.
26. Al-Turjman, F., & Alturjman, S. (2018). Context-sensitive access in industrial internet of things (IIoT) healthcare applications. *IEEE Transactions on Industrial Informatics, 14*(6), 2736–2744.
27. Alabady, S.A., Al-Turjman, F., & Din, S. (2018). A novel security model for cooperative virtual networks in the IoT Era. *International Journal of Parallel Programming, 47*, 1–16.
28. Luhn, H. P. (1960). Computer for verifying numbers. US Patent 2,950,048 (August 23, 1960).
29. Singh, L., & Johari, R. (2015). CLCT: Cross language cipher technique. In *International Symposium on Security in Computing and Communication* (pp. 217–227). Cham: Springer.

Retraction Note to: Context-Aware Location Recommendations for Smart Cities

Akanksha Pal and Abhishek Singh Rathore

Retraction Note to:
Chapter "Context-Aware Location Recommendations for Smart Cities" in: F. Al-Turjman (ed.), *Smart Cities Performability, Cognition, & Security*, EAI/Springer Innovations in Communication and Computing, https://doi.org/10.1007/978-3-030-14718-1_5

The editor has retracted this chapter [1] because a significant part of the text and Fig. 5.3 overlap with a previously published conference paper by Wagih et al. [2]. Akanksha Pal agrees with this retraction. Abhishek Singh Rathore has not responded to correspondence about this retraction.

[1] Pal, A., & Rathore, A. S. (2020). Context-aware location recommendations for smart cities. In: F. Al-Turjman (Ed.), *Smart cities performability, cognition, & security*. EAI/Springer Innovations in Communication and Computing. Cham: Springer.
[2] Wagih, H., Mokhtar, H., & Ghoniemy, S. (2017). Location recommendation based on social trust. In: *Thirteenth international conference on semantics, knowledge and grids (SKG)*, Beijing, pp. 50–55. https://doi.org/10.1109/SKG.2017.00017

The retracted version of this chapter can be found at
https://doi.org/10.1007/978-3-030-14718-1_5

© Springer Nature Switzerland AG 2020
F. Al-Turjman (ed.), *Smart Cities Performability, Cognition, & Security*,
EAI/Springer Innovations in Communication and Computing,
https://doi.org/10.1007/978-3-030-14718-1_12

Index

Printed in the United States
By Bookmasters